A RENAISSANCE GLOBEMAKER'S TOOLBOX

A RENAISSANCE

GLOBEMAKER'S TOOLBOX

Johannes Schöner and the Revolution of Modern Science 1475–1550

JOHN W. HESSLER

Senior Cartographic Reference Specialist, Geography and Map Division, Library of Congress

LIBRARY OF
CONGRESS

The Library of Congress, Washington, DC, in association with D Giles Limited, London

First published in 2013 by GILES
An imprint of D Giles Limited
4 Crescent Stables, 139 Upper Richmond Road,
London SW15 2TN, UK
www.gilesltd.com

Library of Congress Cataloging-in-Publication Data

Hessler, John W..
  A renaissance globemaker's toolbox : Johannes Schöner and the revolution of
modern science, 1475-1550 / John Hessler.
     p. cm.
  ISBN 978-1-907804-16-8 (alk. paper)
  1. Schöner, Johann, 1477-1547. 2. Cartography--History. I. Title.
  GA873.5.S3H47 2012
  526.092--dc23
                        2012023211

ISBN: 978-1-907804-16-8

Partial funding for this book was made possible by a generous
donation from Virginia K. Gray, in memory of Martin Marchbrook Gray.

For The Library of Congress:
Director of Publishing: W. Ralph Eubanks
Editor and Managing Editor: Aimee Hess

For D Giles Limited:
Copy-edited and proof-read by David Rose
Designed by Alfonso Iacurci

Produced by GILES, an imprint of D Giles Limited, London
Printed and bound in China

Front cover: Fig. 44, Celestial globe gores; Fig. 70, Globe representation
of Schöner's *Planispherum*; and Fig. 90, Schöner's *Planispherum*
Back cover: Fig. 43, Celestial globe gores
Frontispiece: Fig. 7, World Map from the 1482 Ulm edition of
Ptolemy's *Geographia*

CONTENTS

FOREWORD

I clearly remember the morning that I walked into the Rare Book Reading Room and found John Hessler hunched over a small volume, intently working out the meaning of a manuscript note in Latin that appeared in the margin of the book he was studying. He looked up and said in so many words, "This place has everything." I am not a specialist in early science, certainly not the arcane world of mathematics and its intersection with astronomy and astrology that Hessler has come to explore. But like so many who have come under the spell of Waldseemüller and his maps, I have over time become more fluent in the field of Northern European science and have begun to better understand what John meant when he marveled at the scope of the collections in the Rare Book and Special Collections Division.

Like so much of our history, the collections go back to Thomas Jefferson and the library of books he sold to the Congress in 1815. This was the first great acquisition the Congress made, and the subjects Jefferson's library encompassed and his system of classification still guides much of the Library of Congress's collection development to this day. In addition to Americana, law, politics, history, literature and the arts, Jefferson's collection includes a strong bias toward the sciences. Of the forty-four chapters in his library classification system, one-third were devoted to the natural and theoretical sciences. Chapters VIII through XV include books in the fields of chemistry, surgery, medicine, animal anatomy, zoology, botany, mineralogy, and the technical arts. Chapter XXV, XXVI and XXVII listed books in mathematics, geometry, and physics. Chapter XXVIII is devoted to astronomy and Chapter XXIX to geography and cartography.

Jefferson's section of books on geography comprises about three hundred titles, or about six percent of his total collection. As might be expected, the largest group of books, about one-hundred eighty titles, relates to the Americas, but there are sixty-two titles on European geography, twenty-four titles on Asia, and thirteen that describe and illustrate the continent of Africa. In addition he has a small group of twenty-five books on classical cartography and a 1620 Paris edition of Strabo's *Rerum geographicarum libri XVII,* a 1678 edition of Dionysius Periegetes' *Orbis descriptio,* and a mid-eighteenth century edition of Pomponius Mela's *De situ orbis.* Jefferson also collected atlases and accounts of voyages of discovery, including copies of the 1595 edition of Ortelius's *Theatrum orbis terrarum,* and the works of Peter Martyr, Giovanni Battista Ramusio, Francesco López de Gómara, Girolamo Benzoni, Theodore De Bry, Richard Hakluyt, Antonio de Herrera y Tordesillas, Fernando Pizarro y Orellana, Baron Lahontan, and William Dampier. Jefferson's collection, with its emphasis on the sciences and geography, became the foundation upon which the Library of Congress built

Figure 2 (detail)

7

its collection of astronomy, cartography, mathematics, and all the natural sciences.

In 1867, about fifty years after Jefferson's collection was purchased by the Library, the collection of Peter Force, comprising some 22,500 books, 37,000 pamphlets, and 300 early maps was purchased for the sum of $100,000. Force was a newspaper man and printer who dedicated much of his time and treasure to purchasing original documents from American history and creating facsimiles for publication. His series was called *American Archives* and over time he managed to publish nine volumes covering the years 1774 to 1776. Of the nearly 60,000 items that came to the Library, Force's collection included 161 incunables and nearly 500 titles printed before 1600. This was the first systematic collection of early printed books to come into the Library's collection.

Force's collection was notable for its emphasis on documenting the evolution of North America from colony to independent state. Like Jefferson he purchased many scientific titles which reflected the new world view that was emerging during the late fifteenth and early sixteenth centuries. Most notably Force's collection included incunable editions of Aristotle, Pomponius Mela, Pliny, Ptolemy, Celsus, Regiomontanus, and a rare edition of Johannes Angelus' *Astrolabium* printed by Ratdolt in 1488. The Force collection also included the 1507 edition of Waldseemüller's *Cosmographiae introductio* printed in Saint-Dié, and Johannes Schöner's *Luculentissima quaedam terrae*, printed in Nuremberg in 1515, as well as editions of the works of Amerigo Vespucci and

Ferdinand Columbus to name only a few of the scores of travel accounts collected.

Force built broadly, purchasing early and sometimes numerous editions of the many important scientific books produced in the late fifteenth and early sixteenth centuries. An examination of the 1868 edition of the *Catalogue of Books Added to the Library of Congress* reveals the scope of his vision and his commitment to the field. Included are editions of the works of Pythagoras, Hippocrates, Theophrastus, Euclid, Archimedes, Eratosthenes, Strabo, Dioscorides, Galen, Apulius Barbarus, Boethius, Gerard of Cremona, Albertus Magnus, Vincent of Beauvais, Paracelsus, Kepler, and Galileo. The list contains many works that were important in rationalization of the sciences in the seventeenth and eighteenth centuries. It brings the field up to date to the middle of the nineteenth century with numerous editions from Europe and the Americas by contemporary scholars.

A few months before the Force collection arrived at the Library, a collection of 40,000 volumes on the history of science was deposited in the Library of Congress collections from the Smithsonian Institution. Much of this material focused on late eighteenth- and nineteenth-century science, with special emphasis on the foreign scientific journal publications. A catalogue of the Smithsonian Deposit was printed in 1866 and lists scientific periodicals from seventeen European countries, as well as Turkey, Africa, Asia, Australia, Polynesia, and the Americas. These journals, filled with both experimental and historical research, have proved to be invaluable source material in all the fields of international science that were developing during the period. It is not a stretch to suggest that the combination of the Force Collection and the Smithsonian Deposit created one of the largest and deepest collections of science in North America, a collection that rivaled those of the universities and colleges of the Northeast that had been growing since long before the Civil War.

The librarian who brought these two large collections into the Library and transformed the Library of Congress into a national center for research was Ainsworth Rand Spofford. He served in various leadership capacities from 1861 to 1908 and was Librarian of Congress from 1864 to 1897. His vision for the development of the Library's collections reflected the influences of the great national libraries of France and Great Britain and their comprehensive approach to building collections. Because of his interest in international trends, he benefited from the developments in library science that were being advocated by German scholars. He was instrumental in making the Library of Congress the "home of copyright" in 1874 and as early as 1876 began urging the construction of a state-of-the-art building. It was not until 1897, when the new Jefferson Building was completed, that this part of his vision was realized. Spofford participated actively in the national discussion about classification, access, cataloguing, and conservation of library materials. More often than not, however, he relied on his own initiatives and those of his associates here at the Library and he established national protocols for bibliographical description and classification that remained the standard for decades to come.

The first significant gift donated to the Library of Congress was the collection of medical books that had been formed by Joseph Meredith Toner. A Washington physician, Toner amassed a collection of 38,000 books, pamphlets, and atlases that documented the history of American science from the colonial period to the Civil War. This collection was augmented by 75,000 manuscript items collected by Toner that supported the medical collection as well as a significant series of manuscripts devoted to the life and times of President George Washington. Toner's collection was accepted for the Library by Spofford in 1882 after authorization from both houses of Congress and the signature of President Chester A. Arthur. With the addition of the Toner Collection, the Library of Congress became one of the great resources for the study of American medicine and medical culture.

For the next twenty years, A. R. Spofford and then the newly appointed Librarian of Congress Henry Putnam systematically built the rare book collections concentrating on Americana, European culture, science and technology, and voyages of discovery. By 1910, when John Boyd Thacher's wife deposited his collection in the Library of Congress, the foundations of the Library collections were set. Thacher's collection contained thirty-four editions of Ptolemy's *Geographia*, hundreds of titles documenting Portuguese and Spanish exploration of the New World, three editions of Waldseemüller's *Cosmographiae introductio*, and nearly 900 incunables documenting the spread of printing in Europe. Thacher also formed a significant collection of books, autographs, pamphlets, and broadsides published during the French Revolutions. Also in 1910, Henry Harrisse, the bibliographer of Christopher Columbus, donated his collection of books to the Library. Harrisse, remembered today as the author of *Bibliotheca Americana Vetustissima,* amassed a significant reference library that documented the cartographic history of the voyages and travels of Christopher Columbus and John Cabot, as well as dozens of other explorers who charted the oceans and the coasts of the New World.

The two most important acquisitions for the rare book collection were to arrive during the early years of the Depression and the middle of World War II. By an act of Congress in 1930, the Library of Congress acquired the Otto Vollbehr Collection of incunabula. Dr. Vollbehr was a German chemist who traveled through the German-speaking territories of Europe after World War I purchasing early printed books and amassing an amazing library of books. The collection contained over 3,000 titles printed between about 1455 and 1501. The Vollbehr Collection included a Gutenberg Bible printed on vellum and books from nearly 650 presses that operated in Europe during the first forty-five years of the printing trade. For the Library's holdings in early science, the acquisition of the Vollbehr Collection was a great boon. In addition to the sacred texts and the classics, this collection included astronomy, mathematics, astrology, natural history, alchemy, cartography, and philosophy, Latin and vernacular translations of Greek and Arabic works on these subjects, as well as early modern scholarship that built upon the scientific knowledge of the ancients. The collection includes obscure early Renaissance treatises, glosses, and commentaries by little-known thinkers who helped set the stage for the scientific revolutions that were to follow in the sixteenth century. This vast collection, much of it by minor writers whose work was saved by good fortune, complements the major works by authors already represented in the Library of Congress collections. Today it is this aspect of the collection that makes it of particular value to historians of science and mathematical cartography.

It was the purchase of the Vollbehr Collection that finally convinced the Library administration that a division responsible for the care and development of the early book collection was necessary. In 1939 the Rare Book and Special Collections Division was formed. All the collections discussed above, as well as many thousands of books from the general collections printed before 1801, came to the new reading room built on the east side of the Thomas Jefferson Building. The first Chief of the Rare Book and Special Collection Division was Frederick R. Goff, whose work during his legendary career as a specialist in incunable printing is one of the great legacies of the world of rare books.

In 1943, Lessing J. Rosenwald created his deed of gift to the Library of Congress for his collection of early printed and illustrated books. It has been written that Rosenwald's gift was the greatest act of cultural philanthropy in American history, a statement strongly supported by many who know the collection. Between the year 1943 and 1978 Mr. Rosenwald purchased important and rare illustrated books which were not already owned by the Library of Congress. Over this period he built a national treasure that has a world-wide constituency. Where the Vollbehr Collection focused on text, the Rosenwald Collection was built to document the development of the graphic arts in the printed book. As a result, Mr. Rosenwald purchased over 1,300 illustrated books printed before 1600, many of which contained scientific illustrations that documented the changes that were occurring in the fields of astronomy, cartography, medicine, mineralogy, chemistry, physics, and mathematics. Rosenwald added 550 illustrated books published between 1601 and 1800, many of which gave graphic evidence of the arrival of scientific research to the modern era.

In addition to these great collections of scientific research that are part of the Rare Book and Special Collections, the Division includes significant collections of gastronomy, food chemistry, agriculture, aeronautics, alchemy, cryptography, engineering, military science, and mechanics, and the collections are still growing. In 2006, with a gift from the Sainsbury Foundation, the Rare Book and Special Collections Division was able to add to our collections two dozen important titles from the history of science that we lacked. In 2008 the Library purchased a copy of Galileo's *Sidereus nuncius,* uncut in original paste paper boards, and just last year a copy of his *Difesa* printed in 1607.

Tracing the growth and evolution of the science collection has been an engaging assignment. Using the resources published by the Library, including Annette Melville's *Special Collections in the Library of Congress* (1980), Leonard Bruno's two volume work *The*

Tradition of Science and *The Tradition of Technology* (1987 and 1995), *The Encyclopedia of the Library of Congress* edited by John V. Cole and Jane Aikin (2005), as well as the collection files, annual reports, and the numerous library catalogues in the Rare Book Division, has guided my path. On a personal level, the process brought me in touch with Jefferson and all those who followed him as benefactors of the Library collections. After nearly two hundred years since the purchase of the Jefferson Library, and the Library of Congress' development into the world's largest collection, I understand what John Hessler meant when he declared, "this place has everything."

Daniel De Simone
Curator of the Lessing J. Rosenwald Collection of the Library of Congress

PREFACE AND ACKNOWLEDGMENTS

Figure 1 (detail)

13

For someone like me, who is more comfortable with an ice ax and a backpack than with a Medieval Latin manuscript, and who would prefer to be traveling off the map, rather than looking at one in a library, the history of Renaissance science and cartography has not been an easy fit. When the Library of Congress purchased the Waldseemüller maps and the *Schöner Sammelband* in 2003 my interests tended toward biogeography and I was at the time researching the fractal properties of the distribution of high-mountain lepidoptera in the Alps of France, a subject far outside the realm of Renaissance studies. Gradually, however, I began to see in the Waldseemüller maps, and in their owner Johannes Schöner, a mystery that needed to be elucidated, and hence a series of books was born. Over the last ten years or so, Waldseemüller and Schöner have given me endless hours of fascination and frustration. They have tested the limits of my command of the Latin language and pushed me to read texts and consult colleagues about questions that I never would have encountered without their ever present image in my mind. This book is the last of my attempts to bring to light the history of these two amazing maps and their preservation by the Nuremberg globemaker and professor of mathematics Johannes Schöner. I hope that what I have written over the years has increased our understanding of these seminal cartographic objects, and I am quite certain that those who follow me will go further than I have been able to.

This book is the first volume in the English language to try to give some context to the life of Johannes Schöner, one of many neglected humanist polymaths of the German Renaissance. It has been my hope that through the telling of his life, which we know much more about than those of Martin Waldseemüller and Mathias Ringmann, some historical frame might be drawn around the very blurry picture that we have of the cartography and science that produced the 1507 and 1516 World Maps.

In the process of writing this, I have benefited from discussions with librarians, curators, rare book and map dealers, historians of science, and cartographers from all over the world, all of whose voices are part of this work and to whom it is dedicated.

It is difficult to thank everyone who helped and supported this book and my other books on the subjects of Waldseemüller and Schöner. First and foremost on this list is Dr. John Hebert, the former Chief of the Geography and Map Division of the Library of Congress. It was John who first excited my interest in these maps and it is due to his help, generosity, and encouragement that my three books on Waldseemüller have come, at last, to be completed. Second, I must acknowledge the support of James Billington, the Librarian of Congress, and a Curatorial Fellowship from the Library of Congress that provided the time and the money for research trips to Vienna and Paris.

Many people at the Library of Congress encouraged this work, including Senior Paper Conservator Heather Wanser, whose interest in the Waldseemüller maps and Johannes Schöner knows no bounds. Dan De Simone, Curator of the Rosenwald Collection, endured hundreds of hours of discussion about the depth of the early science collections at the Library of Congress, and put up with my pointing out its gaps. The comprehensiveness of Dan's knowledge of rare books, and perhaps even more importantly, his dedication and commitment to the art of rare book curatorship and good old-fashioned bibliographic research, are a credit to the field. The Rare Book and Special Collections Division at the Library of Congress is one of the great treasure houses of early science and without the help of its chief, Mark Dimunation, and his small staff of librarians and technicians, who patiently fetched book after book for this project, my work would have been much more difficult.

Besides the staff members of the Library of Congress those of the British Library, the Library of the Paris Observatory, the Vatican Library, and the Österreichische Nationalbibliothek in Vienna have fielded my questions, provided digital images, and helped to track down difficult sources. Special mention must be made of Jay Kislak and Arthur Dunkelman. Arthur, the former curator of the Kislak Collection at the Library, has supported my research on Schöner from its very inception. Jay Kislak purchased the *Schöner Sammelband* after the 1507 map had been removed, and donated it to the Library, thus reuniting the contents of this extremely important Renaissance compilation.

In the three books that I have written on Waldseemüller and Schöner, I have had the great pleasure of working with Ralph Eubanks, Director of Publishing at the Library of Congress, and with Aimee Hess Nash, editor, whose wise council I for one always need and welcome. The enthusiasm of Ralph and of

Dan Giles, the publisher of this book and *The Naming of America*, has pushed me through some of the difficult times, when giving up seemed a better option.

Several scholars have led the way into what is an extremely difficult field and anything I have done here stands in their long shadow. Owen Gingerich helped me with some difficult attributions of Schöner's contorted paleography and generously loaned me his microfilm copies of Schöner's manuscripts in Vienna. Bas Van Frassen, whose books on the philosophy of scientific representation have always provided me with a flexible and formal theoretical framework to think through problems in the history of cartography and science, started me thinking along this path at Princeton many years ago. The late David Woodward, who questioned me concerning the logical structure of a Ptolemaic proof and the role of quasi-periodic functions in the *Almagest*, showed me how careful and patient one must be when approaching these problems.

Finally, I look to the future. Presently, there are several scholars engaged in the pursuit of open questions surrounding Waldseemüller and Schöner, whose talents for this sort of work far exceed mine. I have benefited over the years from long discussions with my co-author of *Seeing the World Anew*, Chet Van Duzer, whose current research into Waldseemüller and the many technical issues regarding its context is second to none. It is determination and patience like his that is necessary for this kind of work, and it is from him that we can look forward to hearing more about this, the greatest of Renaissance mystery stories, in the near future.

PTHOLEMEI

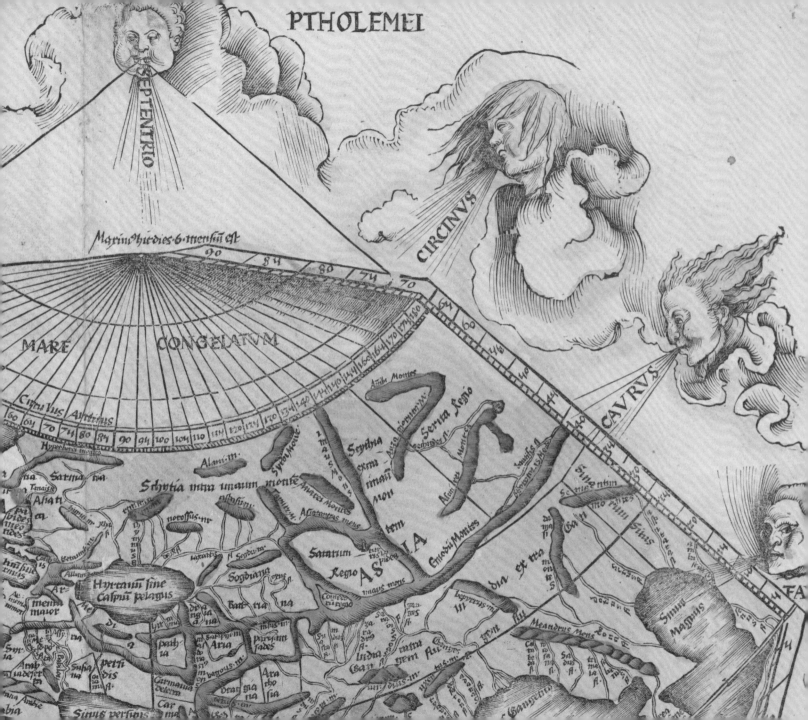

SEPTENTRIO

CIRCINVS

CAVRVS

Marinii hie dies 6. menfiu est

90 84 80 74 70

64 60 44 40

MARE CONGELATVM

Cirtius Austrius

160 64 70 74 80 84 90 94 100 104 110 114 120 124 130 134 140

170 160 154 150 144 140 130 124 120 114 110

44 40

Hyperborea incolae

Sarma na Tanais fl

Alani m

Syebij montes

Alani m Sythia extra ima mon tem

Serica Regio

Sinentn to pun situs

Aliatij Montes

Schytia intra imaum montem

Imaus Montes

Auscei Montes

Ariaca pro mon toriu

Alani mo

Anibei Montes

Emodij Montes

Sinetn Situs

notossus m

Aciacarus mons

Hyrcanij

Iaxartes fl Sogdium

Saturum Regio

imaus mons

dia et ma
montus

Sinarum Situs

Hyrcanii siue Caspii pelagus

Sogdiana

mar ia nus mons

Sinus
Magnus

Bat tria na

Aria Parsij sades

India intra gem

Armenij maior

penti a

Aria

Arabho fia

Meandrus Mons

Sindus fl

Sinus verficus

Sinus persicus

FA

INTRODUCTION

FOR POSTERITY

The Miscellanies of Johannes Schöner

Figure 8 (detail)

But where is this Waldseemüller map? ... Somewhere, in some dark corner of a monastic library, folded away in some oak-bound volume a copy may be sleeping.[1]

— John Boyd Thacher (1896)

ANNO ÆTATIS SVÆ LXIX.

Figure 1
Portrait of Johannes Schöner

This woodcut portrait of Johannes Schöner is one of a number of images of him that survive and comes from the edition of his mathematical works, the *Opera mathematica*, printed four years after his death in 1551.

His life spans one of the most important periods in the history of exploration and science, a time of rapid change and revolution, a time that gave birth to our modern notions of geography, mathematics, and science. He created radical and stunning terrestrial and celestial globes that showed the modern outline of the continents, long before geographers understood their implications. He corresponded with the likes of Nicolaus Copernicus (1473–1543), Bernhard Walther (1430–1504), and Phillip Melanchthon (1497–1560), and in his wisdom he saved two of the Renaissance's greatest masterpieces of cartography from oblivion, the 1507 and 1516 World Maps by Martin Waldseemüller (1470–1520).[2] Besides all this, he published and edited many critical works of mathematics and astronomy of the fifteenth century by no less an astronomer and mathematician than the famous Regiomontanus (1436–1476), and he was partly responsible for the early dissemination and publication of Copernicus' *De revolutionibus,* a book that took a medieval stationary earth and set it in motion about the sun. Yet, most historians of science, of geography, and of cartography have never heard his name and know little of the life and writings of Johannes Schöner (1477–1547).

Johannes Schöner is one of the great neglected personalities of the scientific revolution, standing as he does between the discovery of the New World in 1492 by Columbus and the invention and publication of the heliocentric system of the universe by Nicolaus Copernicus in 1543. Both of these events stunned European society and divided the more conservative religious circles just emerging from the long Middle Ages from the more progressive minds engaged in what we now think of as Renaissance science. Although mostly unknown today, Johannes Schöner was a polymath, a man of many talents, both scientific and artistic, and is one of those rare Renaissance figures who not only lived through these revolutionary times, but whose work, in the form of globes, notes, annotations, and commentaries on these events, have survived the ravages of time and display in stark detail how the new discoveries in geography and astronomy would alter the accepted paradigms of the prevalent world view.

Schöner, although the author of no great scientific work, was a towering figure in the scientific circles of Renaissance Germany, and lived his most productive years in either Bamberg or later as a professor in Nuremberg, when it was one of the most important scientific and artist centers of Renaissance Europe. Rather than a producer of theories, he was instead a disseminator, a compiler, and a transmitter of the new science and mathematics. He seems to have had a keen eye for saving what would later become important to the history of science. His most famous student, Georg Joachim Rheticus (1514–1574), the man who would first announce to the world Copernicus' discoveries in his *Narratio prima,* wrote about Schöner's reputation in a letter to Heinrich Widnauer in 1542. Rheticus recounts, "being attracted by the fame of Johannes Schöner in Nuremberg, who had not only accomplished much in scientific subjects but had excelled in all the best things in life as well".[3]

Figure 2

The *Schöner Sammelband*

The *Schöner Sammelband*, now part of the
collections of the Library of Congress, was bound
together by Johannes Schöner sometime after
1515. The portfolio is arguably one of the most
important collections of Renaissance cartography
and astronomy to have survived. Included in this
one book were the famous 1507 and 1516 World
Maps by Martin Waldseemüller and Mathias
Ringmann, a set of the first celestial patterns for
a globe, a star chart of the southern hemisphere
by Albrecht Dürer, and fragments of two unknown
celestial and terrestrial globes.

Besides being known in his time as a mathematician, astrologer, and globemaker, Schöner was also an avid collector of manuscripts and a bibliophile. Throughout his life he collected the astronomical observations of others, mathematical manuscripts, astrological texts, and fragments from the full range of medieval and early Renaissance scientific and philosophical investigation. It is in this role that he made his mark on history and became known to his contemporaries, and it is through these activities that we will mostly come to know him in this book. Works by Regiomontanus and Georg Peurbach (1423–1461) are just some of the names on the manuscripts he collected, edited, and would later publish.

In his collecting and publishing of the works of others Schöner seems to have displayed an amazing breadth of knowledge of the science of his day. For example, in 1537 Schöner produced an edition of Regiomontanus' *Oration on the Dignity of Mathematics*, which he combined with the Arab astronomers al-Farghana and al-Battani's commentaries on the mathematical sciences.[4] Both of these Arab astronomers were associated with the Maragha observatory located in northwestern Iran. The observatory concerned itself with measuring and updating the observational and physical parameters associated with Ptolemy's *Planetary Hypothesis*, the text of which survives for the most part solely in an Arabic translation.[5]

Schöner's interest in these types of astronomical and geographical observations, taken from across the then known world, is a subject that will loom large in the chapters that follow. We will see him correcting Ptolemy's *Geographia*, updating star charts by Johannes Stabius (1460–1522) and Albrecht Dürer (1471–1528), and annotating his maps, making use of all of these collections in his own experiments as a globemaker and astrologer. Both of these activities were central to Schöner's life and he compiled large collections of latitudes and longitudes of stars and locations of places on earth in their pursuit. These manuscripts became for him a kind of toolbox of sources and information that he would use in the constructions of his globes and astrological predictions. Later, either after he finished, had grown bored, or determined that they were obsolete, he bound them all into sturdy portfolios and books made of leather and wood, thereby preserving them for centuries to come.

One portfolio in particular, a compilation of maps, globe patterns (known as gores), star charts, and fragments of his geographical and celestial labors, now called the *Schöner Sammelband*, or gathering, contained what might arguably be called the most important collection of maps and geographical information to survive from the Renaissance. Into this book, now part of the Kislak Collection of the Library of Congress, Schöner bound two masterpieces of cartography by Martin Waldseemüller, celestial and terrestrial globe patterns of his own design and the earliest printed star chart, by none other than Albrecht Dürer. It is a book whose contents has spurred some of the most hotly contested debates in contemporary historical geography and the history of exploration, containing as it does the only surviving copy of the famous 1507

Figure 3
A sheet of the 1507 World Map of Martin
Waldseemüller in the *Sammelband*

The single sheet of the 1507 World Map showing
the portrait of Amerigo Vespucci before it was
removed from the *Sammelband* for display at
the Library of Congress.

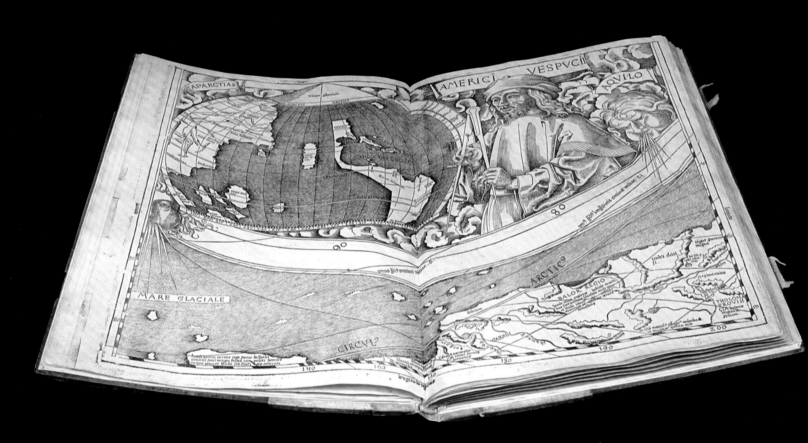

World Map by Martin Waldseemüller, the earliest map to show a Pacific Ocean, dated years before its accepted discovery by Vasco Núñez de Balboa in 1513.

The *Sammelband*, thought lost for centuries, was rediscovered in a small German castle by a Jesuit priest, Father Joseph Fischer (1877–1947), in 1901, and is one of a number of compilations of this type that have survived, most of which are part of Schöner's library in the Österreichische Nationalbibliothek (ÖNB) in Vienna. Fischer was a professor of history at a Jesuit boarding school in Feldkirch, Austria and had spent much his free time traveling around to European libraries looking for information on the voyages and discoveries of the Norse. His trips often took him to small villages and monastery libraries in search of manuscripts, maps, and whatever else he could find in these un-inventoried and out of the way collections.

The story of the re-emergence of Schöner's *Sammelband* begins in 1901, when Fischer received a letter from Father Hermann Hafner, who was the archivist in charge of the collection of books and maps at Wolfegg Castle, ironically located just a few miles from Feldkirch. Fischer arrived at the castle in the summer of that year and began systematically looking through the collection. It was then that, although primarily looking for information on the Norse, he came across a volume that excited his curiosity.[6]

The volume that Fischer paused on contained two of the most mysterious and sought-after maps in the history of cartography, maps whose possible existence had been mythologized by scholars almost since their creation in the village of Saint-Dié in the early sixteenth century. How this volume, which

Figures 4a and 4b
View of Wolfegg Castle and the tower libary
Photograph of the tower library courtesy of Toby Lester

Wolfegg Castle, located in Southern Germany, was the resting place for the *Schöner Sammelband* for over 350 years. Deposited there sometime after Schöner's death in 1547, the compilation remained unknown to scholars until it was rediscovered in the tower library in 1901 by the Jesuit priest Father Joseph Fischer.

FAVONIVS ZEPHIRVS

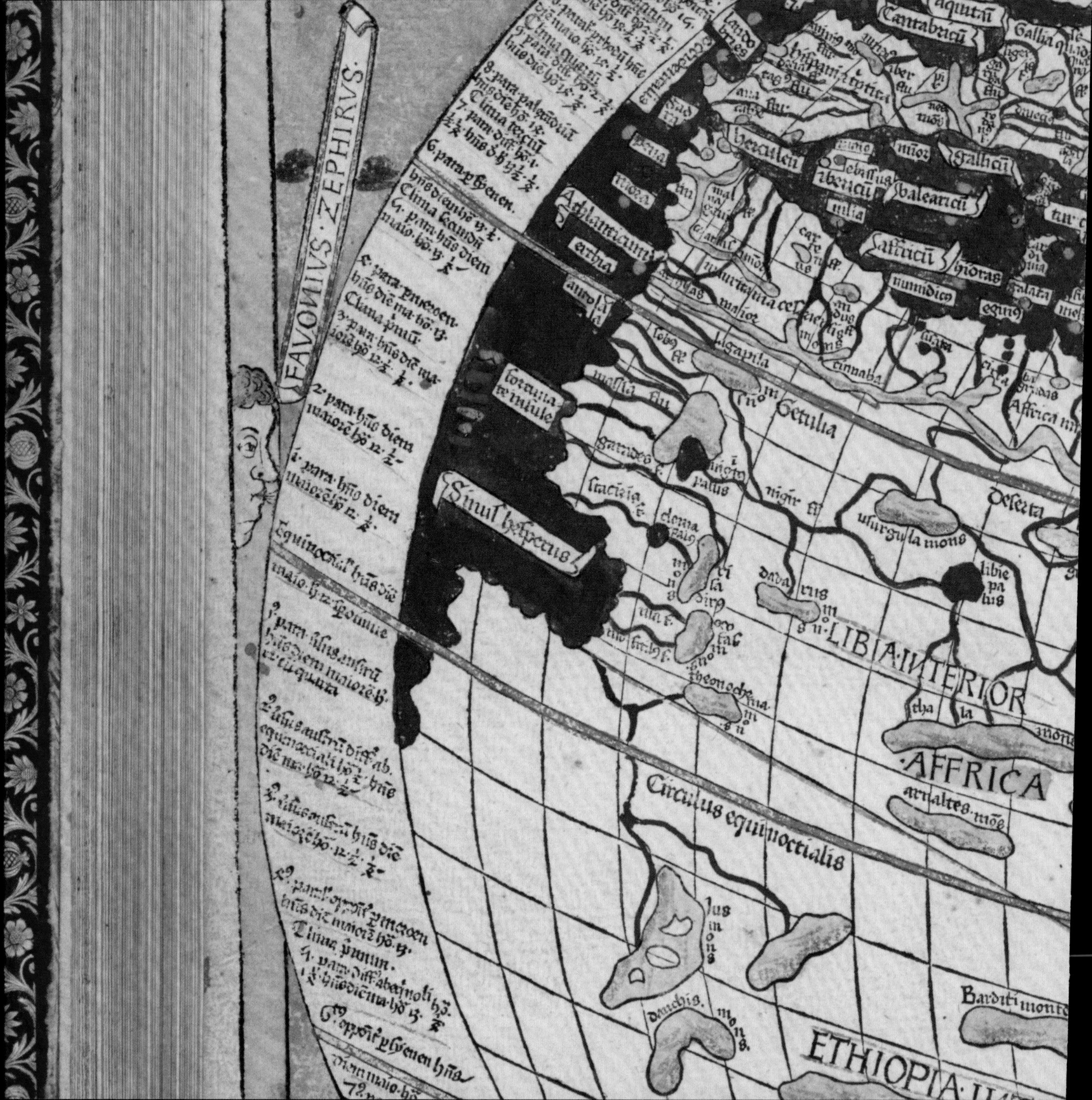

Atlanticum

etha

aurola

fortuia
iufule

Sinus hefpitus

Cantabrici

Gallia

aquitū

hispania

herculeu

Ibericu

Affricu

baleatici

numidici

horas

equor

Getulia

nigir fl.

Deferta

vfargula mons

libie
palus

LIBIA INTERIOR

AFFRICA

arnaltes mõe

Circulus equinoctialis

lus
mons

Barditi monte

dauchis. mons

ETHIOPIA

Fischer quickly concluded contained the only surviving copies of Waldseemüller's 1507 World Map (*Universalis cosmographiae secundum Ptholomaei traditionem et Americi Vespucii aliorumque Iustrationes*) and his 1516 *Carta marina* (*Carta marina navigatoria Portugallen navigationes atque tocius cogniti orbis terre …*), came to be in the castle owned by Prince Franz zu Waldburg-Wolfegg und Waldsee is a mystery that still excites scholars today and is just one of the many questions about Schöner and his compilations that we will consider in this book.

Compilations like the one discovered by Fischer formed the archives and the raw data for Schöner's work as an astronomer, astrologer, and globemaker, and the collections provided him with a kind of toolbox from which he would pull the information necessary for whatever scientific task was at hand. Few archives of this kind survive and the materials his gatherings contain, and which Schöner annotated, provide us today with windows into the mind of one of the most well-informed polymaths of the German Renaissance.

More generally, however, by looking closely at what Schöner thought important enough to preserve in these collections of mathematical, geographical, astrological, and astronomical information, and how he might have utilized it in his work, we will gain deep insights into the epistemological revolutions that occurred at the end of the fifteenth century and the beginning of the sixteenth. The years between the amazing discoveries of Columbus and Copernicus saw the beginnings of the birth of modern scientific thought and in the chapters that follow we will see Schöner fully engaged in the intellectual transitions

from the science of Aristotle and the geography of Ptolemy, to that of Copernicus and Ferdinand Magellan (1480–1521). The dates of the materials that Schöner compiled are mostly from 1475 to 1540 and represent a cross-section of central scientific materials from pre-Copernican science. Each of these books and manuscripts is interesting in its own right but taken together they provide a case study for the use and transmission of scientific information in the Renaissance through the eyes of a contemporary consumer of these materials. Schöner's toolboxes are nothing short of encyclopedic and his use of them helps us understand in a unique way how our modern scientific world view came into being.

Johannes Schöner was born on January 16th, 1477 in Lower Franconia. Unfortunately we know very little of his early life. At some point, however, he became very interested in science and mathematics and began to acquire books on those subjects along with many on geography and cartography. One of his earliest purchases was a copy of an *Ephemerides* calculated by the fifteenth-century astronomer Regiomontanus. An ephemerides is typically a list of the positions of astronomical objects and is used for both observational astronomy and astrology, subjects that Schöner was interested in throughout his life. The title comes from the Greek word *ephemeris* and literally means diary. In the case of this particular copy, Schöner took the Greek meaning quite literally and began to keep a journal of some of his activities in the margins.

The diary and the *Ephemerides*, preserved in the Österreichische Nationalbibliothek in Vienna,[7] tell us little about Schöner's intellectual pursuits at the

OPERA

Mathematica

IOANNIS SCHONERI
CAROLOSTADII IN VNVM VOLVMEN
CONGESTA, ET PVBLICAE VTILITATI
studiosorum omnium, ac celebri famæ Norici
nominis dicata.

Impressa Norinbergæ, in officina Ioannis Montani
& Vlrici Neuberi, Anno Domini
M, D, LI.

Figure 5

Title page from the 1551 edition of
Schöner's *Opera mathematica*

Schöner wrote a great deal during his
lifetime and published the works of many
other mathematicians and astronomers.
This is reflected in this edition of his
mathematical works, which includes many
original texts, but also several that are
directly derived from the manuscripts that
he collected.

time, but give us some hints about his personal life. In the margins of the *Ephemerides* is his account of a relationship that he began with a woman, named Kungigunde Holocher, in 1499. Schöner tells us that she bore him three children, one in 1502, another in 1503 and a third in 1504, all while he was a Catholic priest, having been ordained in the bishopric of Bamberg in 1500. Previous to becoming a priest Schöner matriculated at the University of Erfurt, probably from the winter term of 1494 through sometime in 1498. The diary tells us that he was appointed vicar in his hometown of Karlstad on June 4th, 1504, with the information the diary contains breaking off in 1506, when he was again back in Bamberg.

From this point on most of the information we have about Schöner comes from his works, his globes, and some of the volumes that he edited and published. Many of the prefaces that he wrote for the edited works are especially useful and from them we know that he was appointed, after converting to Protestantism, the first professor of mathematics at Nuremberg in 1526, a post he held until his death in 1547.

Throughout his life Schöner collected books and manuscripts and bound them together like the *Sammelband,* and his library contains a great many of the texts that one would want if one was an astronomer or geographer in the early sixteenth century. The list of works found in his *Opera mathematica*,[8] published in numerous editions after his death, gives some idea of the range of his interests, encompassing everything from elementary mathematics to the use and construction of complex astronomical instruments.

After Schöner's death the contents of his library passed into the hands of the merchant Georg Fugger (d. 1569) and from him it was handed down to Fugger's son Phillip Eduard (1546–1618) and then to his great grandson Albert III (1624–1682). The entire contents of Fugger's library containing some 22,000 books and manuscripts, mostly on mathematics and astronomy, was purchased in 1656 by Emperor Ferdinand III, and currently resides in the Österreichische Nationalbibliothek.[9] One of the many mysteries surrounding Schöner's *Sammelband* is how that one volume became separated from the mass of material purchased by Ferdinand and made its way to the castle at Wolfegg.

Looking at the contents of the *Opera mathematica*, together with the materials found in the Library of Congress and the ÖNB, it is clear that Schöner read, annotated, and copied notes from the full range of Renaissance geography, mathematics, and astronomy texts from the late fifteenth and early sixteenth centuries. His holdings included many titles in multiple copies and editions.

In the field of geography and cartography Schöner owned copies of the 1482 Ulm and the 1513 Waldseemüller editions of Ptolemy's *Geographia*. We know from an annotation in his copy of the 1482 *Geographia* that he purchased it in the fall of 1507, the very year of the printing of the famous Waldseemüller map that would become such an important source for Schöner's terrestrial globes.[10] The 1482 edition was updated in 1513 by Waldseemüller and Mathias Ringmann, who not only added Greek place names to the volume, but also provided a series of modern

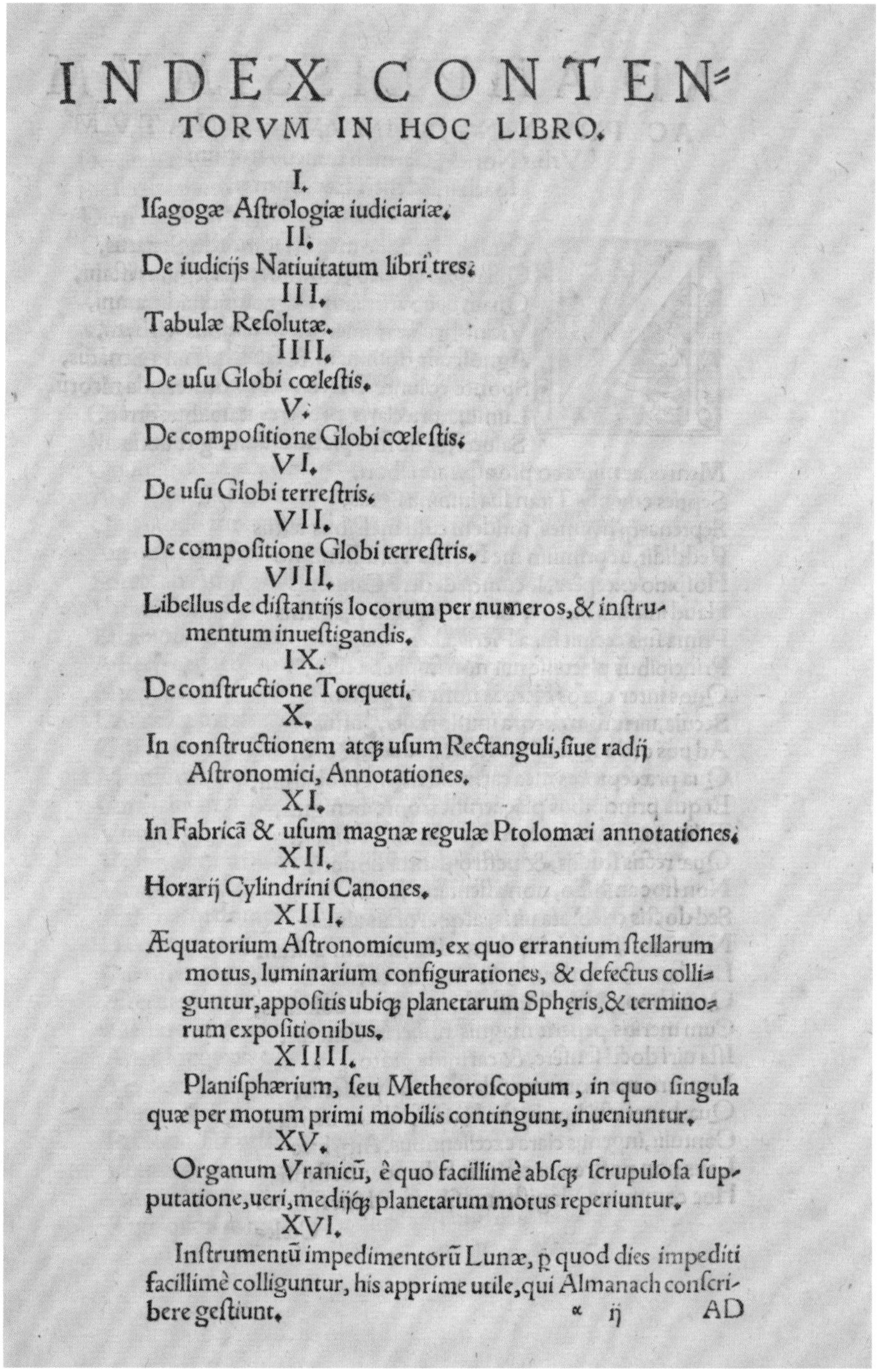
INDEX CONTEN-
TORVM IN HOC LIBRO.

I.
Iſagogæ Aſtrologiæ iudiciariæ.
II.
De iudicijs Natiuitatum libri tres.
III.
Tabulæ Reſolutæ.
IIII.
De uſu Globi cœleſtis.
V.
De compoſitione Globi cœleſtis.
VI.
De uſu Globi terreſtris.
VII.
De compoſitione Globi terreſtris.
VIII.
Libellus de diſtantijs locorum per numeros, & inſtrumentum inueſtigandis.
IX.
De conſtructione Torqueti.
X.
In conſtructionem atq; uſum Rectanguli, ſiue radij Aſtronomici, Annotationes.
XI.
In Fabricã & uſum magnæ regulæ Ptolomæi annotationes.
XII.
Horarij Cylindrini Canones.
XIII.
Æquatorium Aſtronomicum, ex quo errantium ſtellarum motus, luminarium configurationes, & defectus colliguntur, appoſitis ubiq; planetarum Sphęris, & terminorum expoſitionibus.
XIIII.
Planiſphærium, ſeu Metheoroſcopium, in quo ſingula quæ per motum primi mobilis contingunt, inueniuntur.
XV.
Organum Vranicũ, è quo facillimè abſq; ſcrupuloſa ſupputatione, ueri, medijq; planetarum motus reperiuntur.
XVI.
Inſtrumentũ impedimentorũ Lunæ, p quod dies impediti facillimè colliguntur, his apprime utile, qui Almanach conſcribere geſtiunt.

α ij AD

Figure 6
Contents of the *Opera mathematica*

The contents of the *Opera mathematica* reveal the depth and variety of the intellectual pursuits of Johannes Schöner. Titles ranging from elementary mathematics to very complex natal astrology held his interest throughout his lifetime. Schoner was a polymath, equally at home in the study of the geography of the New World and the new astronomy of Nicolaus Copernicus.

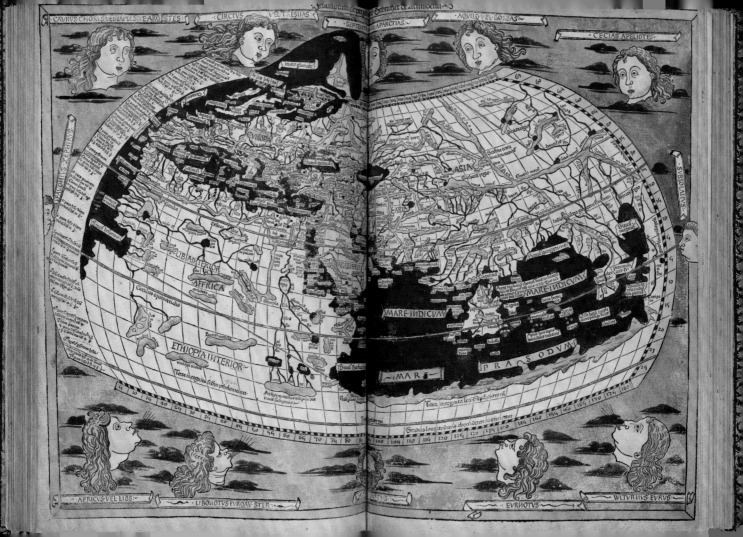

Figure 7

orld Map from the 1482 Ulm edition of
olemy's *Geographia*

e influence of Ptolemy's *Geographia* during the
naissance in Europe cannot be overstated. Rescued
m antiquity, the text was published in many
itions by the turn of the sixteenth century, with
s edition, produced in Ulm in 1482, being one of the
ost beautiful of the early editions. This was the book
at Schöner himself owned and annotated.

artin Waldseemüller and Mathias Ringmann
oduced an edition of the *Geographia* in 1513 that
as to be extremely influential in the years that
lowed. The two mapmakers from Saint-Dié
dated Ptolemy's maps and coordinates using
eek manuscripts, producing a book that contained
t only the Ptolemaic image shown here, but also
dated images like the World Map in Figure 9.

Figure 9 (detail)

ASIA

Auranthis Regio

Sena Regio

Regio

Scithia intra imaum

India intra Gangem Regio

ganges fl.

India est Gange

Tap bana

Sinus Gan geticus

Mallaqua

Berzigingipani

Regiu bac

Senla

INDICVM

Tangut prouina

Signi pura

Cathem

Thebeta priura

Manin mariaa

India suzior

Cynika oui na

quirirum

Sirus Magnus

Murbili xeauira

Maabar Regnu

Regnu nar

Loir Regnu

Iaua Maior

Regnu

Auua

Iaua Minor

Figure 9

Admiral's chart from the 1513
Waldseemüller *Geographia*

This updated World Map was included
by Waldseemüller and Ringmann in
their 1513 edition and provides a contrast
to the other World Map they included,
shown in Figure 8.

maps that showed portions of the New World. In the
chapters that follow we will investigate how Schöner
read these influential geographic texts and how
he used them in the creation of his globes, which,
inspired by Waldseemüller's radical revision of the
shape of the continents, also present a revolutionary
view of the layout of the continents and the new
discoveries of Columbus and Vespucci.

In his astrological and astronomical writings
Schöner would utilize the sources of his day in
unique ways. Throughout his later years he seems
to have spent a great deal of time reading books like
Regiomontanus' *Epitome of the Almagest of Ptolemy*,

Georg Peurbach's *New Theory of the Planets*, and
Ptolemy's *Tetrabiblos*. The two astrology books that
Schöner wrote in the late 1530s will be of particular
interest to us in the chapters that follow as they
give us reason to believe that Schöner's major
interests resided in astrology. It is possible that his
compilations of information regarding the positions
of stars, and of the latitude and longitude of cities and
towns, found in books like the *Geographia,* and also
on the maps he collected and annotated, served this
larger purpose.

In addition to annotating the seminal scientific
treatises mentioned above, Schöner also took extensive
notes on astronomical and geographical themes
that cannot easily be traced to their original sources.
For example, he wrote and copied a 'geographical
miscellany' containing jottings on things like map
projections, distance measurement, and how to convert
different latitude and longitude grids. One work in the
miscellany, entitled *Regionum sive civitatum distantiae,*
gives directions for making a terrestrial globe from
a map and would have been especially important to
Schöner's early globemaking activities.[11]

Although Schöner used most of his books for
immediate informational purposes, he also was
intent on preserving them for future generations.
In the *Sammelband* and in his copy of the 1482
Geographia, Schöner affixed a bookplate that bears
the inscription, "Schöner gives this to you posterity;
as long as it exists there is a monument to his spirit."
The words on the bookplate imply that Schöner
made a conscious effort to save these compilations
for future generations, and suggests that he knew

This bookplate was pasted by Schöner into his books and survives in several of his extant compilations. This one, found in the *Sammelband*, gives us a clear idea of Schöner's purpose in saving the manuscripts and maps that he bound into them. The text translates, "Schöner gives this to you posterity; as long as it exists there is a monument to his spirit."

something of the gravity of the situation to come, as the Reformation, Counter-Reformation, and the Papacy would more and more seek to control the flow and content of scientific information, culminating a century later in the Galileo affair.[12]

One need not look too far to gain some measure of assurance about Schöner's attitude toward the dissemination and preservation of great scientific works. In the preface to his edition of Regiomontanus' *On Triangles of Every Sort* (*De triangulis omnimodis*), published in 1533, and which is dedicated to the "Most esteemed and Learned Lord Senators of Noricum," Schöner explains his reasons for preserving for future generations important mathematical and geographical books and information. Schöner begins the preface by praising the astronomer Regiomontanus and the strength of his learning,

> At long last this work has been discovered, to the increasing glory of your city. As it has come into our hands, so it must be made available for the profitable study of all. None of his works— however inornate, imperfect, or even discarded as they may seem—should be considered other than great … But as his *Index* shows the work he had planned here was extensive. So it is certainly desirable that the monuments of his labors and of others be preserved.[13]

The passage refers to an *Index*, which is an ambitious list of medieval and classical texts in mathematics, geography, and astronomy that Regiomontanus had hoped to edit and publish in his

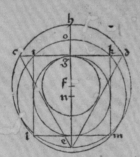

Figure 11
Title page from the *De triangulis* by Regiomontanus

Of all of the books by Regiomontanus that Schöner published, his book on triangles was probably the most influential. The book updated the methods of trigonometry and provided a textbook for the foundations of mathematics needed by anyone who wanted to study astronomy. The book was critical to the education of Copernicus, and was used throughout the sixteenth century.

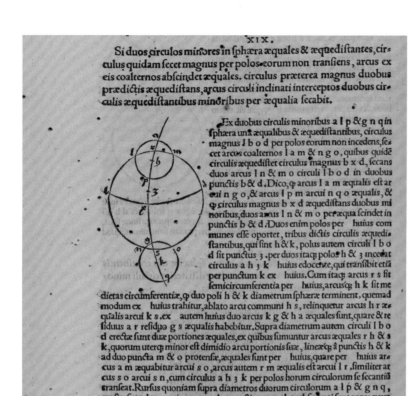

Figure 12

Example of spherical trigonometry in the *De triangulis*

This page gives an example of the type of
mathematics found in Regiomontanus' text. Spherical
trigonometry was something needed by anyone
wanting to understand the astronomical models of
Ptolemy as found in the *Almagest*.

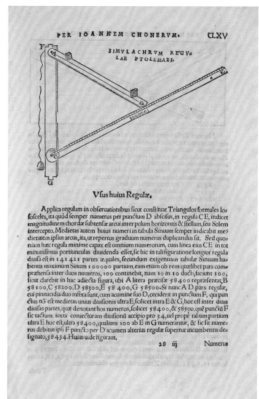

Figure 13

Great Rule of Ptolemy

Throughout his life Schöner was interested in astrological
instruments and this is reflected by the number of descriptions of
them found in his *Opera mathematica*. Before the early seventeenth
century all astrological observation was accomplished with the
naked eye and these types of instruments were necessary in order
to quantify difficult celestial measurements.

Figure 14

Title page to the *Algorithmus demonstratus*

Schöner published this book on elementary mathematics and calculation in 1534, based on notes left by Regiomontanus on a thirteenth-century mathematician named Gernardus. Nothing is known of the mathematician, but the book is a guide to arithmetic calculation and was published in many forms and under many different titles during Schöner's lifetime.

lifetime. A student of Georg Peurbach, Regiomontanus, whose real name was Johannes Müller, died in 1476 at the age of forty, long before he could get to even a fraction of the titles he listed. Most of his completed works would only be published later by Schöner, including *On Sines and Chords*, and *Descriptions and Explanations*, which contains details on the make-up and operation of astronomical equipment like the Torquet, and the great rule of Ptolemy.

Regiomontanus was a central figure in astronomy and mathematics and his book on triangles marked a new phase in the development of mathematics in the West. At the age of eleven, Regiomontanus entered the University of Leipzig and later lived and studied in Vienna from 1450 to 1461. During his early years in Vienna it was a center for the study of mathematics and astronomy and it was at that time that he met Georg Peurbach. Besides his textbook on planetary astronomy, Peurbach also worked on mathematics, composing a book on sines and chords which was influenced by an earlier work of elementary mathematics called the *Algorithmus* by John of Gmunden (before 1385–1442). Regiomontanus studied these mathematical texts in detail and copied most of their contents into what has become known as the Vienna *Rechenbuch*, the manuscript of which remains extant in Johannes Schöner's library.[14]

Besides his interest in elementary mathematics and trigonometry, Regiomontanus also developed a deep understanding of algebra, which was something quite new at the time of his investigations.[15] In 1463, Regiomontanus came across a manuscript of the Greek writer Diophantus and the next year, in his lectures in Padua in 1464, he praised the algebra found in it. Regiomontanus' notes on Diophantus unfortunately do not survive, but he did manage to copy what has been identified as a thirteenth-century treatise on arithmetic and algebra by an unknown medieval mathematician named Gernardus. The manuscript, entitled *Algorithmus demonstratus*, was the basis for a book of the same title that Schöner published in 1534.

Schöner's respect for the work of Regiomontanus was not his only reason for editing and preserving great works of science and geography, for he understood that there were other things afoot in Reformation Germany. He continues in his preface to *De triangulis*, calling his period "turbulent times," and writes that "we are hemmed in by the stupidity of men, and we see all the arts nearly abandoned by all mortals." He looks upon his work of publishing and editing as something only the future will appreciate.

> For this reason no one understands the praise I may evoke by the famous works I publish. For what is more worthy of the praise of the republic of letters than their protection? And in what battle do the arts seem more freely to spend their strength than in praise of culture and its defense? …
>
> You know the times. No one really looks for a rebirth of the arts. They are so silent and neglected, it may be feared that the idiots will wipe them out. God alone can preserve them by inspiring the leaders of the city to water the seeds of knowledge and nurture the seedlings that sprout.[16]

So far we have talked here mostly of Schöner as a geographer and as an astronomer working in fields whose conceptual foundations we might recognize today. Even though this is only partially true, it is important in the chapters that follow to remember that the early sixteenth century had not yet seen the revolution in science that overthrew some of the older, less empirical notions of cause and effect, mostly derived from Aristotle. Experimental and observational methods had not yet blossomed into the forms that they would become only a century later, with the writings of Francis Bacon and the observations of Galileo. Astronomy for Schöner, although still an inherently observational and predictive science, was still very much attached to astrology and his longest and most detailed writings surround a specific form of the subject: natal astrological prediction.

Renaissance astrology was divided into four different parts: prognostication, nativities, elections, and interrogations. The first of these, prognostications, was uncontroversial for the most part and concerned itself with the prediction of things like natural disasters, plagues, and weather. The other three, however, would come in for criticism as they all concerned predictions about the future of individuals.

Natal astrology, the form that held Schöner's foremost interest, is based on concepts that assert that every individual's path in life and their personality can be determined by constructing a natal chart for the exact date, time, and location of their birth. In September of 1545, Schöner would publish his *Three Books on the Judgment of the Nativities*,

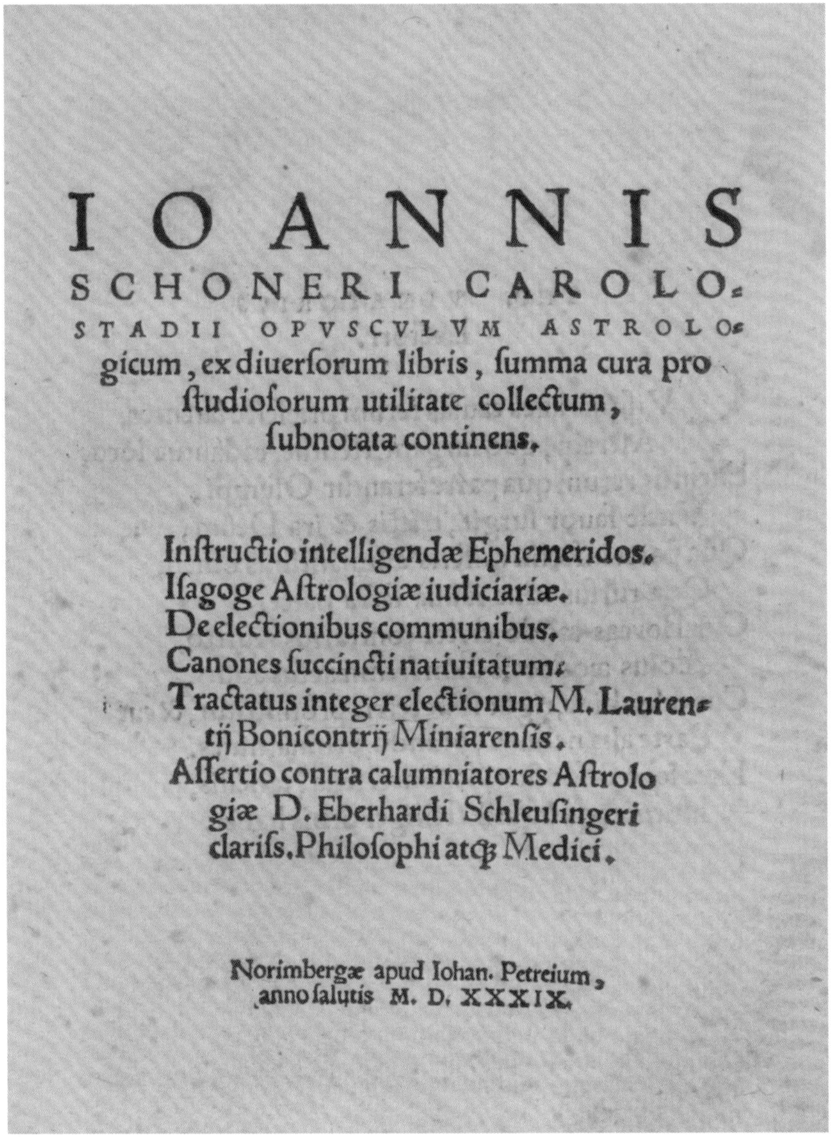

IOANNIS
SCHONERI CAROLO.
STADII OPVSCVLVM ASTROLO-
gicum, ex diuerſorum libris, ſumma cura pro
ſtudioſorum utilitate collectum,
ſubnotata continens.

Inſtructio intelligendæ Ephemeridos.
Iſagoge Aſtrologiæ iudiciariæ.
De electionibus communibus.
Canones ſuccincti natiuitatum.
Tractatus integer electionum M. Lauren-
trij Bonicontrij Miniarenſis.
Aſſertio contra calumniatores Aſtrolo
giæ D. Eberhardi Schleuſingeri
clariſſ. Philoſophi atϙ Medici.

Norimbergæ apud Iohan. Petreium,
anno ſalutis M. D. XXXIX.

Figure 15

Title page to Schöner's *Introduction to Astrology*

One of the least studied parts of Schöner's work is that of his astrology. He wrote two books on the subject at the end of the 1530s that try to update the methodology of Ptolemy and his medieval commentators.

Figure 16
Schöner's Pillar Dial

The pillar dial is a traveler's instrument designed to allow its user to tell the hour of day by the angle of the sun at any location at any time of the year. This image shows a pattern for a pillar dial, which is cylindrical. The pattern shown here would have needed to be wrapped around a tube for it to work properly.

which like much of his work, is an attempt to pass on information that he thought was critical for future generations. The contents of Schöner's text, especially the first book, parallel Ptolemy's *Tetrabiblos* or *Liber quadripartitus*, which is the most complete astrological text that has come down to us from antiquity. Schöner presents a very medieval conception of natal astrology, with little in the way of updated information, and it is uncertain what edition of the Ptolemy he worked from as there were many Latin manuscripts of the text circulating at the time in addition to the first printed edition from 1484.

The entire text of the book uses a single natal chart, cast by Schöner, for the Habsburg Emperor Maximilian I. Schöner uses this one example to go step-by-step through the methods employed to make an accurate natal chart. Considering the number of surviving manuscripts, Nuremberg seems to have been a hub of natal astrology during Schöner's lifetime. Most of Nuremberg's famous artistic and scientific residents would have had horoscopes cast and, because it was necessary to actually write down these horoscopes on natal charts in order to interpret them, a great many survive.

Reading Schöner's *Nativities* one cannot help but be drawn to speculate on why Schöner found it so important to have the latest maps and star charts in his possession. Schöner's instructions for making these charts are extremely clear and show that it is the very information contained on updated maps and star charts that is necessary in the accurate production of a natal chart. It is in his astrology that all of Schöner's interests seem to come together and that his

annotations of geographical texts, his updating of star charts, and his preservation of maps, seem to make the most sense.

One of the least studied aspects of Schöner's work is his interest in instruments used for both astronomical and astrological observations. Many plans for astronomical instruments appear in his *Opera mathematica,* most of which are based on the descriptions of Regiomontanus. Two examples of these are the *Torquetum* and the *Pillar Dial.*

A pillar dial is also known as a cylindrical dial, a shepherd's dial, or a *horologium viatorum.*[17] The instrument is generally of cylindrical shape, so the plan shown here is typically wrapped around a small cylinder. Along the bottom of the instrument is a grid representing the path of the sun along the zodiac for the year, together with a calendar scale. The curving lines that span the body of the instrument show the curving arc of the sun's shadow at the same hours at different times of the year. If this schematic plan of Schöner's were actually made into an instrument one can see the apparent motion of the sun through the zodiac from summer to winter, and then in the reverse direction from winter to summer.[18] The instrument was used to determine the latitude of the traveler as well as to find the hour and angle of the sun. This would allow, at least in principle, for the owner of the dial to figure the hour in whatever location he might find himself.

A second instrument Schöner writes about, and one that he actually built, is the torquetum. In a letter to Willibald Pirkheimer from January of 1524, Schöner explains how he built a torquetum and taught himself its operation, later using it to observe the position and movement of Halley's Comet, which returned in 1531.[19] The torquetum is a complicated instrument, whose origins go back to the thirteenth century and it was first described by Franco of Poland in 1284.[20]

Regiomontanus, whose description Schöner's no doubt derives from, associated the torquetum with the twelfth-century astronomer Jabir ibn Aflah, who lived in Seville. The actual use of the instrument is still a matter of debate in scholarly circles. Most likely the instrument was designed for use in Ptolemaic astronomical observation, which demands a precise determination of the position of a star in terms of its ecliptical latitude and longitude.

Looking at the large breadth of his interests and his writings, Schöner will present us in this book with some categorical difficulties as we attempt to place his work within the context of Renaissance science. Treating the full diversity of his work, from mathematics to astronomy, from astrology to cartography, is problematic from both a practical and methodological point of view. In the person of Johannes Schöner the full range of Renaissance science is on display, only a small part of which has been the subject of scholarly attention.[21] Most of what has been treated centers around the disciplines that are considered orthodox—natural philosophy, medicine, and mathematics. This is a selection of subjects that vastly restricts the culture of debate taking place within the circles of natural historians and philosophers of the period. What we find in the writings of Schöner, and in the writings of others, is that much of the most

innovative epistemological reasoning about what it means to be a science are coming from the more applied and "unorthodox" sciences.[22] Many of those sciences such as astrology, alchemy, cartography, and natural magic are in many cases yet to receive adequate scholarly treatment historically, with even less being known about their individual methodologies and logical development.

We have therefore, in Schöner and in his *Sammelband*, the perfect chance to investigate how older scientific paradigms mixed with newer forms and began the process of scientific and epistemological growth that produced our modern world. In the notes of an astronomer, of a mathematician, and of an astrologer, we will in the chapters that follow see the birth and infancy of our modern scientific world, and how it formed through the eyes of a single man who witnessed its most important revolutions. It is by looking closely at how the discovery of new worlds, both terrestrial and celestial, affected the intellectual development and interests of a single well-informed and scientifically inclined humanist, that we can begin to understand more clearly the paradigm shifts that put us on the path to our modern scientific world view.

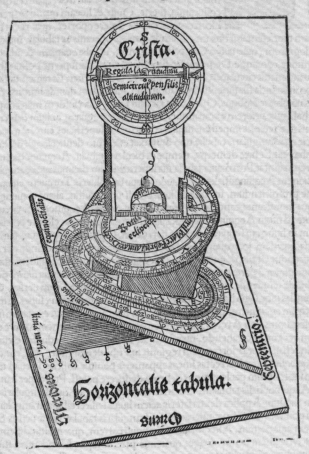

IOANNIS SCHO-
NERI CAROLOSTADII, FRANCI,
MATHEMATICI, DE CONSTRVCTIONE
Torqueti, Dogmata sex.

Crista.

Regula latitudinum

Semicirculi pensilis altitudinum.

Zodiacus eclipticae

æquinoctialis

linea meri.

Septentrio.

Horizontalis tabula.

Meridies

DOGMA I.

Duas quadrangulares tabulas, unamq; circularem unà cum
triangulo eleuationis Æquinoctialis, pro constructio-
ne Torqueti præparare.

27 ♄   Primum

Figure 17

Torquetum

The torquetum is an instrument whose use
is not well understood by scholars today.
According to Schöner, the instrument is used
for determining the precise position of a star
in terms of its elliptical coordinates. In his
correspondence we learn that he used an
example of one to track what would become
known as Halley's Comet, in 1531.

THE *SCHÖNER SAMMELBAND*

At the Library of Congress

Figure 22 (detail)

1507 WORLD MAP

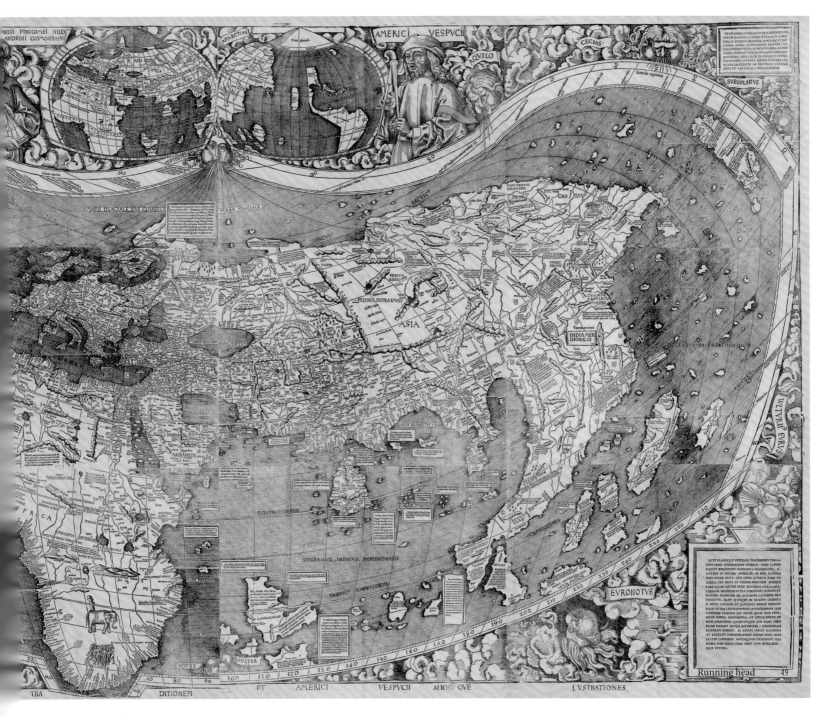

Figure 18
Sheet 1

The four corner sheets on the 1507 map all have text blocks that are important for understanding the map. This one mentions Vespucci, who is pictured in sheet 3, and Columbus as both having a role in the discovery of the New World. The text translates:

> Many have regarded as an invention the language of the famous poet that "beyond the stars lies a land, far away from the path of the year and the sun, where Atlas, who holds up the heavens, revolves on his shoulders the axis of the world, set with shining stars." For there is a land, discovered by Columbus, as captain of the King of Castile, and by Amerigo Vespucci, both of them of great ability, which, though a great part of it is found beneath, "the path of the year and the sun" and between the tropics, nevertheless extends near 19 degrees beyond the Tropic of Capricorn toward the Antarctic Pole, "beyond the path of the year and the sun." In this region more gold has been found than any other metal.

Figure 19
Sheet 2

A portrait of Ptolemy dominates the sheet, with a small inset map of the Old World. The *Geographia* of Ptolemy was one of the most important geographical sources for Waldseemüller's map. On the inset one can see a line marked *terra Ptolemeo incognita* (lands unknown to Ptolemy), which Waldseemüller used to divide the world that Ptolemy knew from that which had only been recently discovered. The large block of text is taken from the Roman geographer Pomponius Mela and discusses whether or not the entire lands of the earth are surrounded by sea.

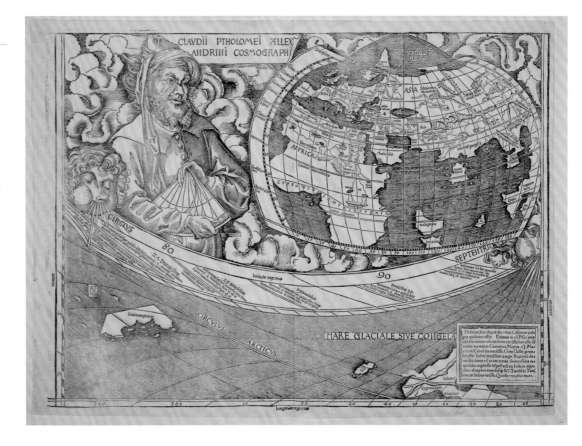

Figure 18 (detail)

Figure 20
Sheet 3

The majority of this sheet is taken up with a portrait of Amerigo Vespucci, after whom Waldseemüller named the new continent America (although he described it as an island). The inset map shows this New World and the Pacific Ocean, along with the Far East beyond. If one looks closely over the shoulder of Vespucci (above the letter Q in *AQVILO*), one can see either a wasp or a fly, placed there either to symbolize Vespucci, as the Latin for wasp, *vespa*, recalls his name, or if a fly, as a talisman that printers commonly used to protect their work from damage. The artist and printmaker Albrecht Dürer employed a realistic representation of the fly in some of his paintings and woodcuts, in the hope that the viewer would swat at the insect, thinking it real—thereby validating Dürer's talent as an artist.

Figure 21
Sheet 4

Here we see the land of Cathay (China), most of which is represented in Marco Polo's account of his travels. Waldseemüller gives a short explanation of his reason for making the map and describes its contents in the large text block found in the corner of the sheet. It translates:

> In representing the overall appearance of the entire world, it has seemed right to place on the map discoveries of the ancients, and to add what has since been discovered by the moderns, for example the land of Cathay, so that anyone who is interested in such matters and who wishes to discover various things may see their wishes fulfilled and be grateful to us for our labor, when they see everything that had been discovered here and there, or recently explored, carefully and clearly placed together, so that it may be seen with merely a glance.

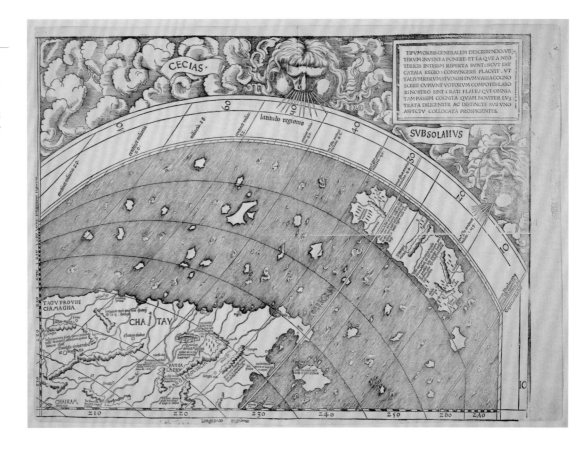

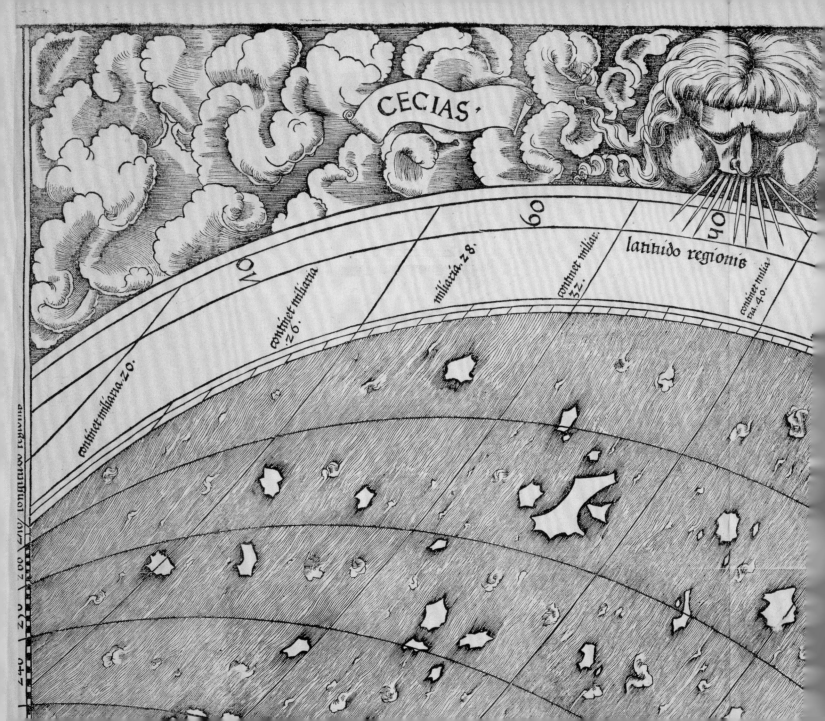

CECIAS·

longitudo tabulæ / 617 / 00 7 / 2 / 440

continet miliaria. 20.

continet miliaria. 26.

miliaria. 28.

continet miliar. 32.

latitudo regionis

continet miliaria. 40.

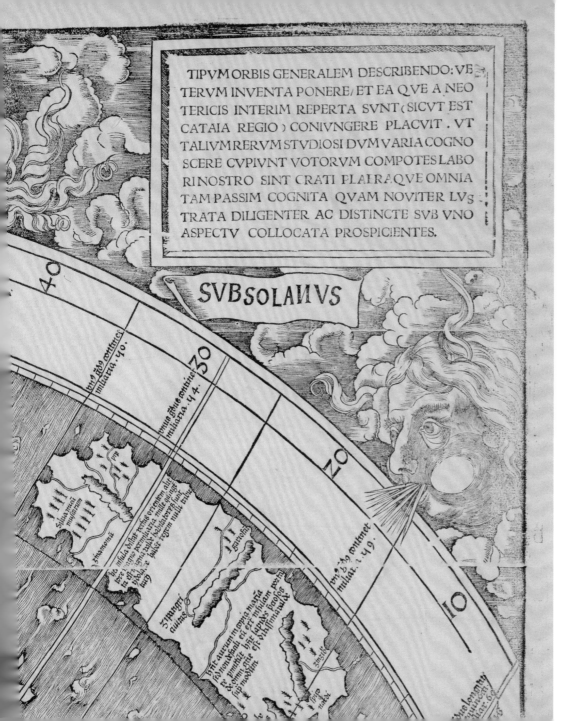

Figure 21 (detail)

TIPVM ORBIS GENERALEM DESCRIBENDO: VE
TERVM INVENTA PONERE, ET EA QVE A NEO
TERICIS INTERIM REPERTA SVNT (SICVT EST
CATAIA REGIO) CONIVNGERE PLACVIT . VT
TALIVM RERVM STVDIOSI DVM VARIA COGNO
SCERE CVPIVNT VOTORVM COMPOTES LABO
RI NOSTRO SINT CRATI FLAI RAQVE OMNIA
TAM PASSIM COGNITA QVAM NOVITER LVg
TRATA DILIGENTER AC DISTINCTE SVB VNO
ASPECTV COLLOCATA PROSPICIENTES.

SVBSOLANVS

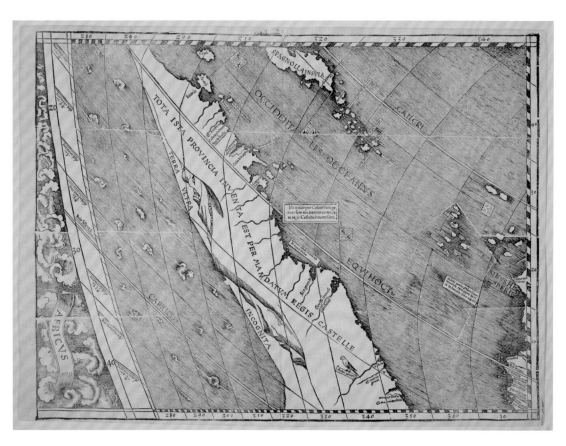

Figure 22
Sheet 5

This sheet shows the top of South America and part of the Pacific Ocean. Much has been made of the western coastline of the new continent on the map and its polygonal or linear structure. Several scholars have speculated that Waldseemüller might have known more than the historical record has told us about this west coast. The Pacific Ocean would not be discovered until 1513, and the end of the continent was not rounded until 1522, leading to questions regarding how Waldseemüller might have known about the western ocean. Schöner in his terrestrial globes clearly shows this western ocean.

Figure 23

Sheet 6

The Mediterranean and the north of Africa on the Waldseemüller 1507 map are very much Ptolemaic representations. The majority of the shapes and outlines of the coastlines of these regions are similar in appearance to those found in Ptolemy's *Geographia,* even though Waldseemüller would have known them to be quite different. The tip of Africa near Gibraltar is especially obvious in this regard, showing the horn-like appearance found on much earlier Ptolemaic maps. In doing this, Waldseemüller appears to be trying to emphasize the contrast between the old and new geography that he is displaying on the map as a whole.

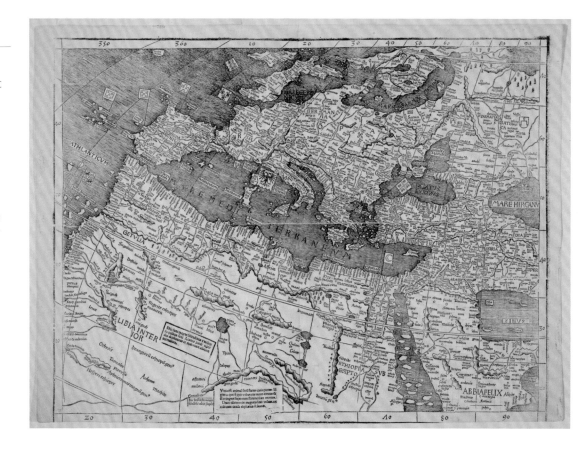

Figure 23 (detail)

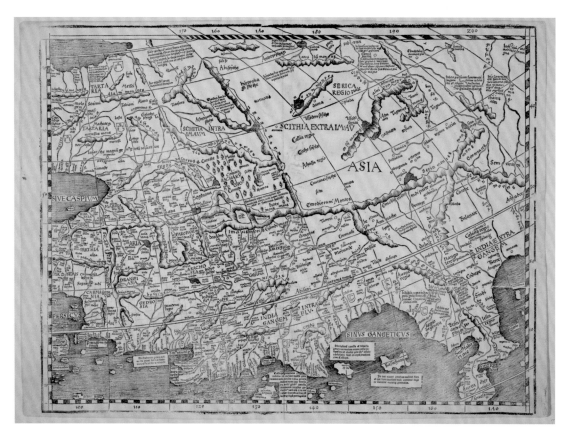

Figure 24
Sheet 7

This sheet shows an area that must have interested Johannes Schöner a great deal. In this region we find that Schöner annotated the map with a red grid, which he most likely used to extract coordinates for his globes or for his astrological activities.

Figure 25
Sheet 8

The Far East and China are extensively annotated with small text blocks, most of which come from the travels of Marco Polo. One can also observe, if one looks closely at the sheet, that it is peppered with small crosses inside shields. These are the symbol of the mysterious Prester John, a mythical figure about whom various tales were told concerning his rule over a lost Christian nation, trapped amidst Muslims and pagans in the Middle and Far East. Waldseemüller discusses these symbols and many of the others found on the map in a long explanatory caption in his *Cosmographiae introductio*, which was written as a sort of guidebook for the 1507 World Map. In the book Waldseemüller explains that he divided the map using symbols and flags based on the rulers of particular regions. He also explains that these flags were colored. Sadly, no color remains on the flags found on the only surviving copy of the 1507 map.

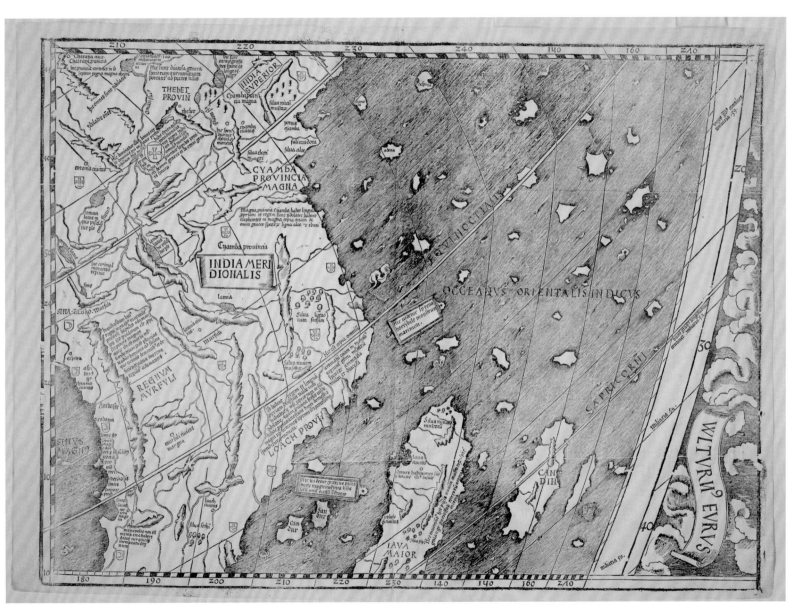

Figure 24 (detail)

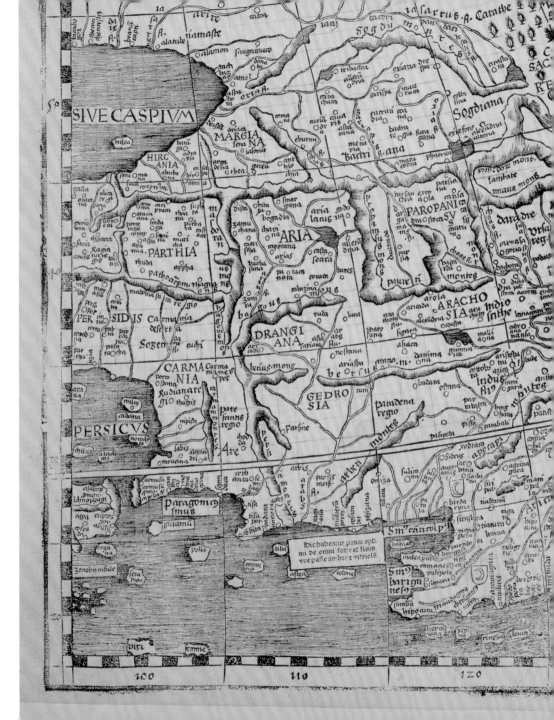

ASIA

Athasia regio · · · · · balliana · ote · · · Sem · · · ani m
Cauranei Scythe · trauta · abrogana · · Serin mon · · · · · Se
sota · · · orosana · · · · ma
tchorne · · · · · · · · · · · · · · · mdi parte asanamana · · · · · Catope · · · mon tes
monsturris · · · · · · · archinara · · · · · · thadam arrorra · · · Athadre
lapidis · · · · · · · · · · · · · · · · · · · Ser
bilthe · maus · Emodiorum Montes · Emodij mons · · · · · · · · · · · · · · · · · · ius
mons · · · · · · · bermge · · · · · · salatha · · ·
· · Tatorei · salam · · · · · · · da · · · · · · · · Balanate
mon & Chilndi mare · · · pura · · · · · · · · · · · · · · msi · · · · ·
· mont · · · · Calatis reg
Gangane · · · · · · · · · be eldana ·
· · · · · · · · · · · pi · Ammathe ·
Patho · Coranoli · · · · · · · · · r · · · · · Cuadia inbac · · · · · · · edislagra · · · · · spiberig
clarareg · dialropem · · · · · Dabise · · · · · ·
· · · · · · · · · · gan · · · · Morunde · · · breuosi · Passale · · · · · thalabachinu · · · · · possma · · · · · INDIAE XTRA
· ra · Nango · · · · GANGEM
· · · · · · · · · · malitha · lette es puismula · romam
Caspiria · · Biolnige · · palm · · · · · · astena gurum · Tilede · alasongr · · · · · · · · · m e · · · · · · · · · · · Barre
· · · · · · · · · bothu · · ux · · · · · · · · · · trlidra · · · · · · · · · a · · · · · · · · · · Anagara · Codupe
· · · · · · · · · · · · Astagbura iins · · · · · · mame · · · · · · · · n ·
Adisati · urentius mons · · · · · · orco · · na · · · · · Aurea · · · · · · · · · · · · · · · Sudi
· · · · · · · · · · · panasta · · · · · · · · · agmin · · · · · · regio ·
· · · · · · · · · · · · · · · gorza ·
· · · · · · · · · · Obirgu ·
INDIA · sat9 INTRA · lesta
GANGEM · · · ms GEM FLV · Regio · Besin · antro · · regio
· argentea · · phagu · · ·
· SINVS GANGETICVS · · · · · · · · · · · · · · · · · · Sin9 Saba
· ricus
· · · · · · · · · · · Hic nascunt canella et diuersa ·
· · · · · · · · · · · genera aromatum eciam perle pi · Sin9 pimu
Smus aga · · · · stant hic et aliofar ges isti9 insule · hia9
rius · · · · · · · · ydololatre sunt et coytu tractant ·
· · · · · · · · · · · · cum Calliqut. ·
· Balathata · · · · · · · · · · · · · · · · · · Aurea · · cher
· · · · · · · · · · · · · mora ·
Calliqut prouincia · · · · · rchim · · · In bac dicunt conches multas fieri · · · · · · · · · · · · · · · · · ·
nobilis in ea sunt multa gene · · · · · · · · et qui banc incolunt nudi continue dege · · · · · · · · · · · · · · · ·
ra minerarum pimeta z alia genera · · · · · · re dicuntur vocariq gimnatas. · · · · · · · · sonelsatinh
mercatorum que veniut · armadiom ·
de nulti ptib9 canella ·
zinimonia zinziber ga · Sissem · lusam ·
riofol sandalu z de om
nibus speciebus: hec est
iucta p rege portugallie

130 · · · · · 140 · · · · · 150 · · · · · 160 · · · · · 170

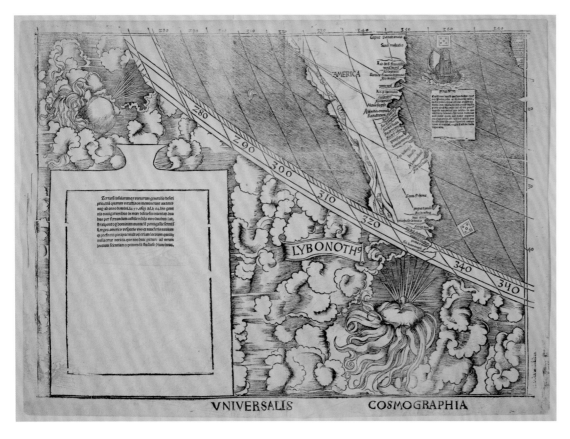

Figure 26
Sheet 9

The famous 'America' sheet shows the location that Waldseemüller chose to place his homage to Vespucci on the new continent. The name America is one that Schöner would also employ. This sheet also contains one of the four corner text blocks giving details of the explorers who discovered the new lands, and contrasts the limited knowledge of the ancients with Waldseemüller's geography. The Latin text translates:

> An overall delineation of the various lands and islands, including some of which the ancients make no mention discovered lately between 1497 and 1504 in four voyages over the seas, two commanded by Fernando of Castile, and two by Manuel of Portugal, most serene monarchs, with Amerigo Vespucci as one of the navigators and officers of the fleet; and especially a delineation of places that were not known previously. All this we have carefully represented on the map in order to provide true and accurate geographical knowledge.

Figure 27

Sheet 10

This region of Africa was unknown to Ptolemy; hence Waldseemüller's representation is based on new geographic knowledge. This is most apparent if we look closely at the place names that dot the coastline, almost all of which are based on Portuguese sources rather than Ptolemy's Greek.

On this sheet we find the only annotation, besides the red lines, by Johannes Schöner. Chet Van Duzer has traced the source of this annotation to a globe of Martin Behaim's (1459–1507), from ca. 1492. The text, written in Schöner's hand, describes the group of islands off the African coast and reads:

> The islands of St. Martin. There it is summer when here it is winter. All of the birds and animals of these lands are of a different sort [than ours]. The zibethus [that is, civet] is called algalia by many; among us it is called muscus.

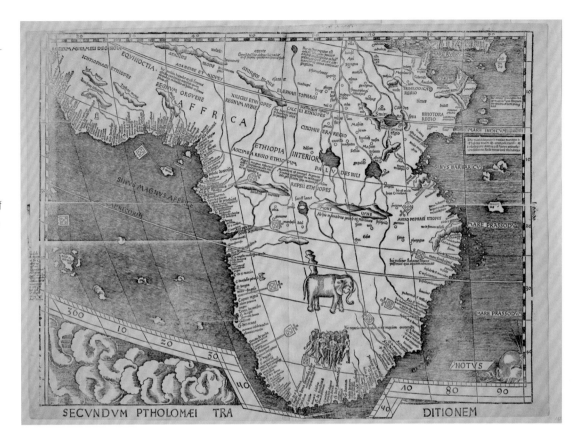

Figure 27 (detail)

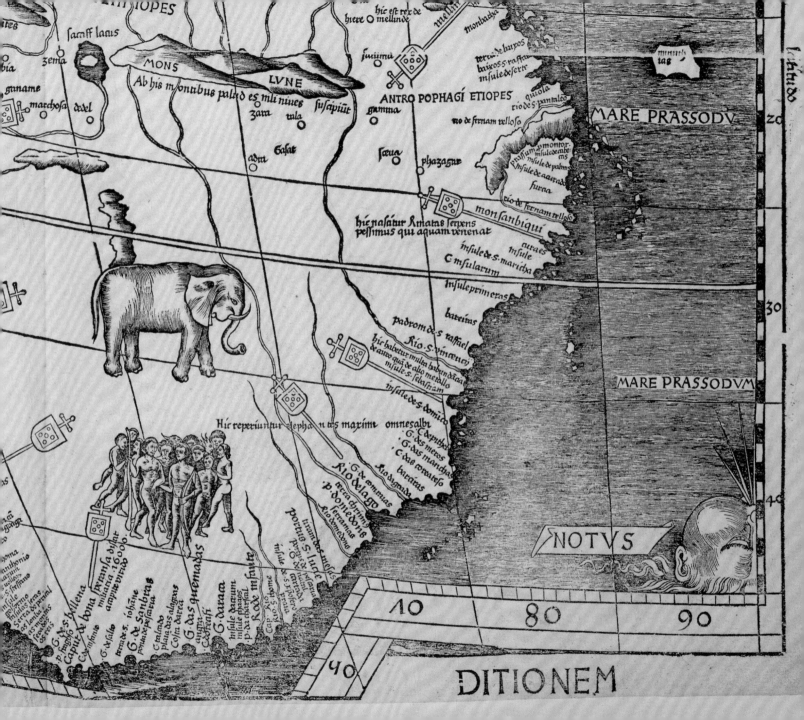

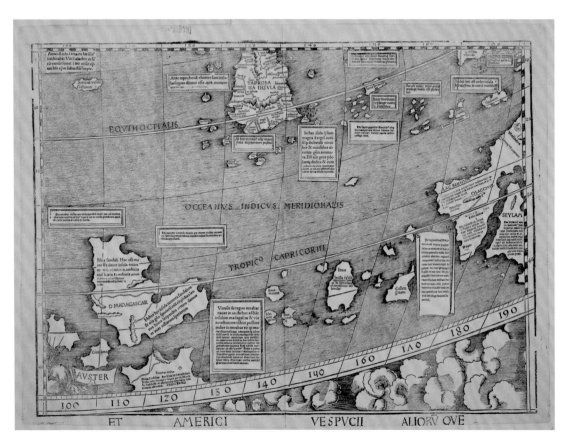

Figure 28
Sheet 11

The sea of islands shown on this sheet is studded with small text blocks describing some of their features. The islands in the north on the sheet are derived from Ptolemy, while those in the south once again come from the travels of Marco Polo, as does the small landmass in the far northwest corner. This sheet reflects in stark detail the many sources Waldseemüller needed to draw from in order to create the 1507 World Map. Sources such as the Martellus map, the Caverio chart, and Ptolemy's *Geographia* were not commonly available in his time, nor were many exploration and travel narratives. It highlights one of the great unresolved problems in the history of early cartography: that of the transmission of geographic source material, something Schöner certainly would have recognized.

Figure 29
Sheet 12

Of all the text found on the 1507 map, this is the largest single example. It has also proved problematic to the dating of this only surviving copy of the map. Unlike some of the other text blocks, this one is pasted onto the map and not printed directly onto the sheet itself. The paper on which the text was printed also has printing on the back that has been found to come from a palm-reading manual whose earliest known edition dates from 1515 (*Ein schones Buchlin der Kunst Chiromantia*).

The text found in this block is one of the few places where Waldseemüller reveals to us that he knew he was doing something revolutionary with his World Map. In the last line, he explains to readers that if they are inexperienced with the new geography, they should not condemn what they see presented to them on the map. Rather, they should be patient because *this is how they will come to view their world*. The full text translates:

> Even though most of the ancients were interested in marking out the extent of the world, many things remained unknown to them; for instance, in the west, America, named after its discoverer and that is now known to be a fourth part of the world. Another, to the south, a large part of Africa, which begins about seven degrees this side of Cancer and extends in a great expanse southward, well past the torrid zone and the Tropic of Capricorn. A third example, found in the east, is the land of Cathay and all of southern India, past one hundred and eighty degrees of longitude. All of these places we have added to the ones that were known before, so that those who are interested and love knowledge of this kind may see all that is known to men in the present day, and that they may approve of our work. This one request we have to make, that those who may not be acquainted with geography shall not condemn all that they see before them until they have learned what will surely be apparent to them later on, when they have come to understand what they see.

Figure 29 (detail)

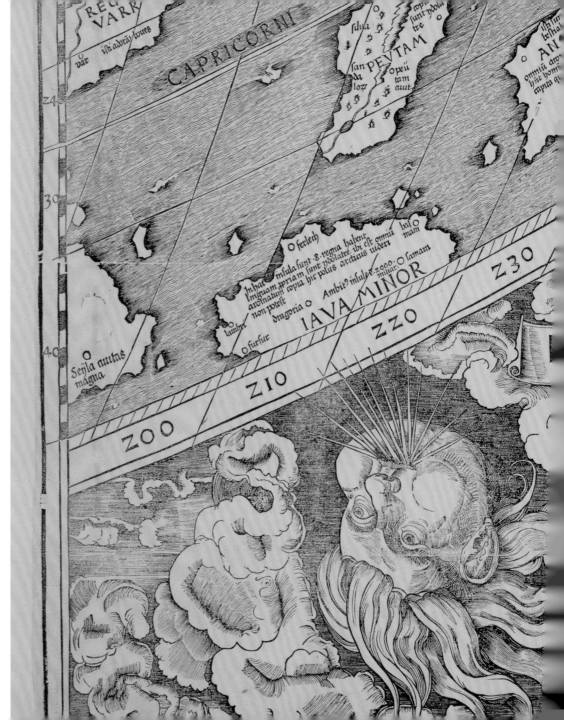

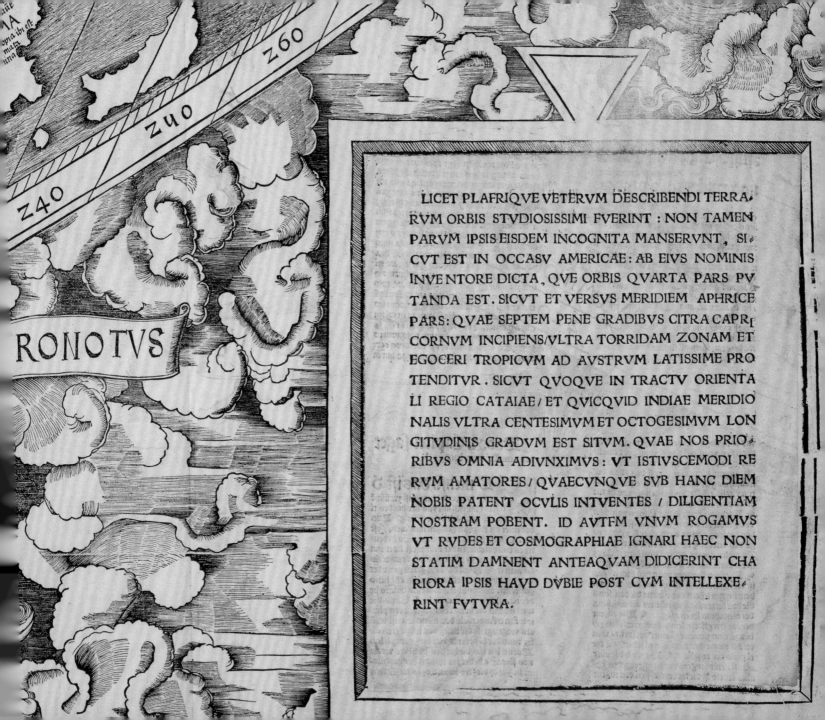

Z 60

Z 40

Z 40

RONOTVS

LICET PLAERIQVE VETERVM DESCRIBENDI TERRA
RVM ORBIS STVDIOSISSIMI FVERINT : NON TAMEN
PARVM IPSIS EISDEM INCOGNITA MANSERVNT, SI
CVT EST IN OCCASV AMERICAE : AB EIVS NOMINIS
INVENTORE DICTA, QVE ORBIS QVARTA PARS PV
TANDA EST. SICVT ET VERSVS MERIDIEM APHRICE
PARS : QVAE SEPTEM PENE GRADIBVS CITRA CAPRI
CORNVM INCIPIENS/VLTRA TORRIDAM ZONAM ET
EGOCERI TROPICVM AD AVSTRVM LATISSIME PRO
TENDITVR. SICVT QVOQVE IN TRACTV ORIENTA
LI REGIO CATAIAE / ET QVICQVID INDIAE MERIDIO
NALIS VLTRA CENTESIMVM ET OCTOGESIMVM LON
GITVDINIS GRADVM EST SITVM. QVAE NOS PRIO
RIBVS OMNIA ADIVNXIMVS : VT ISTIVSCEMODI RE
RVM AMATORES / QVAECVNQVE SVB HANC DIEM
NOBIS PATENT OCVLIS INTVENTES / DILIGENTIAM
NOSTRAM POBENT. ID AVTEM VNVM ROGAMVS
VT RVDES ET COSMOGRAPHIAE IGNARI HAEC NON
STATIM DAMNENT ANTEAQVAM DIDICERINT CHA
RIORA IPSIS HAVD DVBIE POST CVM INTELLEXE
RINT FVTVRA.

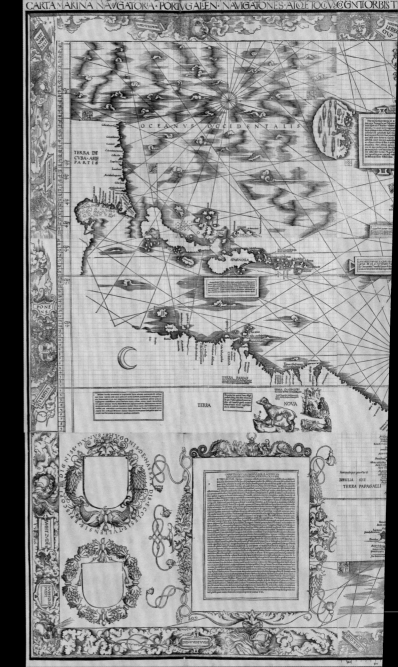

1516 *CARTA MARINA*

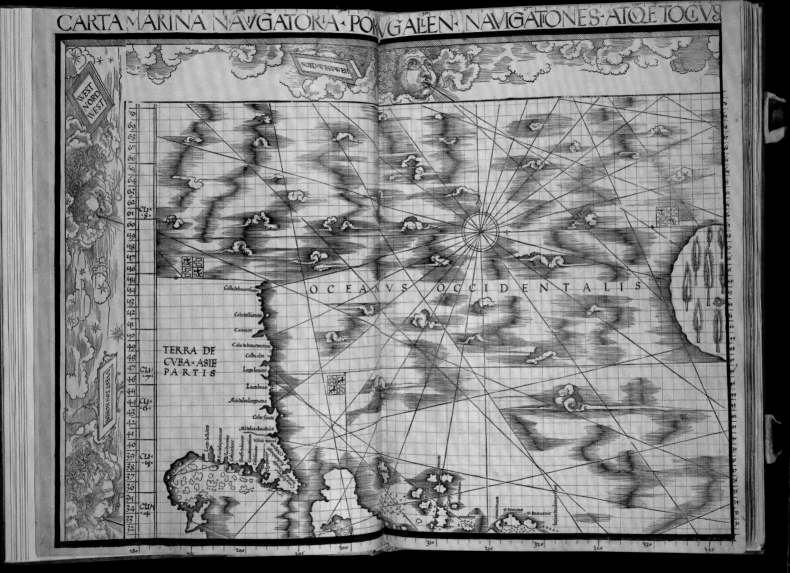

NORDWESTPWEST

WEST NORD WEST

WESTEN NORDE

OCEANVS OCCIDENTALIS

Costa del mar sbaso

Cabo del liante

Cannor

Cabo de bona ventura

TERRA DE CVBA·ASIE PARTIS

Costa alta

Laga lamoro

Lacabras

Asio de los lagatos

Cabo santo

Rio de las almadias

Figure 30
Sheet 1

This is one of many sheets on the *Carta marina* that Schöner annotated with a scale in the left-hand margin. The sheet is important in that it contains a retraction of Waldseemüller's 1507 World Map and makes the New World a part of Asia.

Figure 31
Sheet 2

The *Carta marina* is very different from the 1507 World Map, being made in the form of a Portolan, or sailing chart and in the fact that it is populated with graphic representations of cities and animals of all sorts. What Schöner would have made of these differences is unclear but he seems to have been mainly interested in the coordinates on the map as represented by his hand-drawn latitude scale and red grid.

NORD WEST
LEBEC
CIVS

70
69
68
67
66
65
64
63
62
61
60
59
58
57
56
55
54
53
52
51
50
49
48
47
46
45
44
43
42
41
40
39
38
37

G.bas vingharfus

TERRA NOVA

Seßo das Rocellas

GRIPPAT

C.dabal

TERRA LA
BORATORIS

Hec terra Corterati inuenta est ex man
dato regis Portugallie per Casparum Cor
terati capitaneu.in duorum nauium anno
domini.1501.quam ob sui magnitudinem
litoris plusquam 600. miliarin protenden
tis firmam esse opinabatur.habet hec plura
litate magnoru fluminu. habet gentem
qae huius. habet domos ex maximis lignis
cōstructis.quarū tectora ex coriametis pisci
um cōpacta, vestesqz oru sunt de pellibus fera
rū.quarū in estate pilos ab extra.in hieme
vero ab intra vertere portat. Sunt signati fa
cie.tanquā indi.circet ferro et loco illius.lapi
deis instrumetis vtuntur. Magnā habet co
piā lignoru de genere pini.etiā multos pis
ces Salmones Alecca & Strumulos.

ILHA

acsmrea

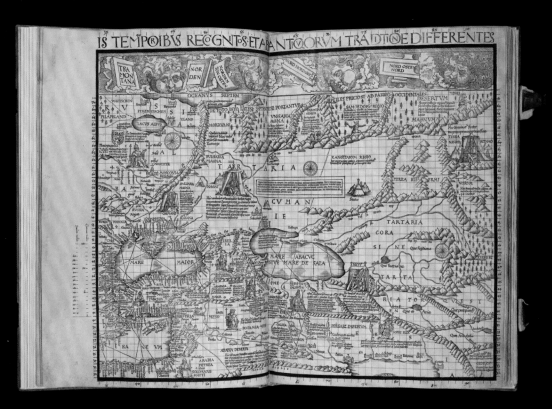

Figure 32
Sheet 3

Continuing his population of the map with
monstrous creatures and cultural symbols,
Waldseemüller crowds this sheet with monsters
from three different races, each enclosed by
mountainous regions.

Figure 33
Sheet 4

Schöner's annotated grid covers this area running
into the border and out to his hand-written scale on
the map. The sheet is dominated by Waldseemüller's
characterization of the Great Khan.

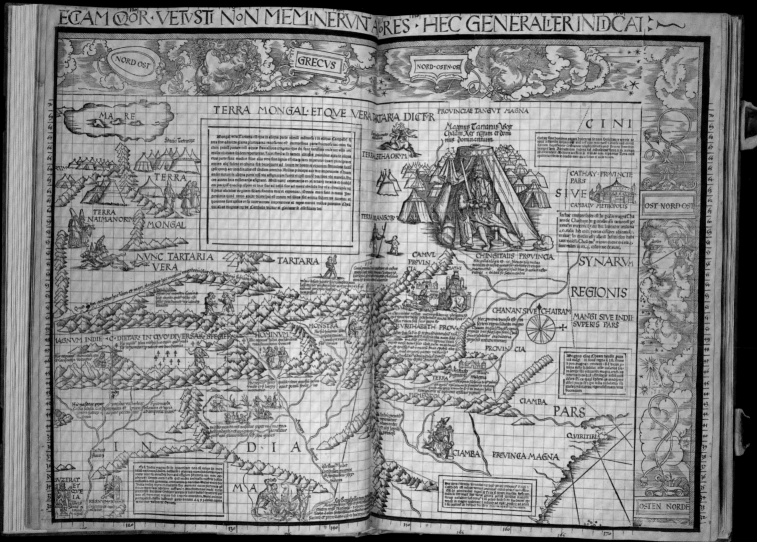

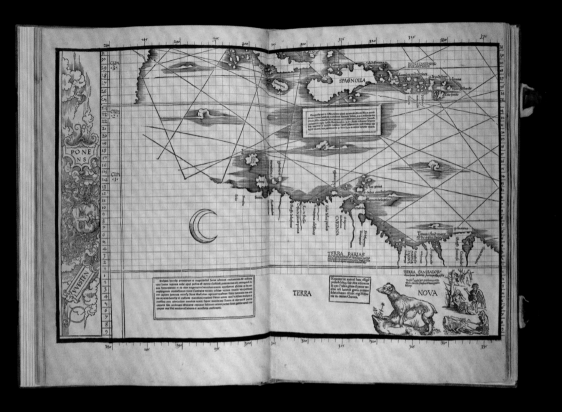

Figure 34
Sheet 5

This sheet has two scales, one printed on the left-hand side, and another added by Johannes Schöner. On the 1507 map this geographic area was part of the continent of South America and had a distinct eastern coastline, which disappears from the 1516 map.

Figures 35 and 36 (page 84)
Sheet 6 and Schöner's manuscript

This sheet shows the coast of Africa and presents a bit of a mystery. Both the printed sheet 6 and the manuscript version of it, drawn in the hand of Johannes Schöner, were found in the *Sammelband*. The manuscript sheet, however, was the only one that was actually attached to the binding. The printed version was simply tipped into the book. Whether or not Schöner had access to the printed sheet remains an open question, but the Renaissance historian Chet Van Duzer has postulated that the manuscript sheet was made in preparation for one of Schöner's later globes from the 1520s.

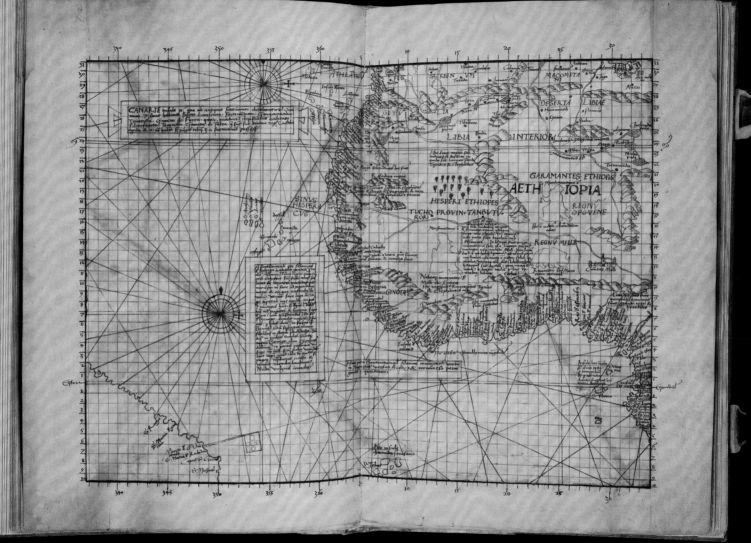

Figure 36

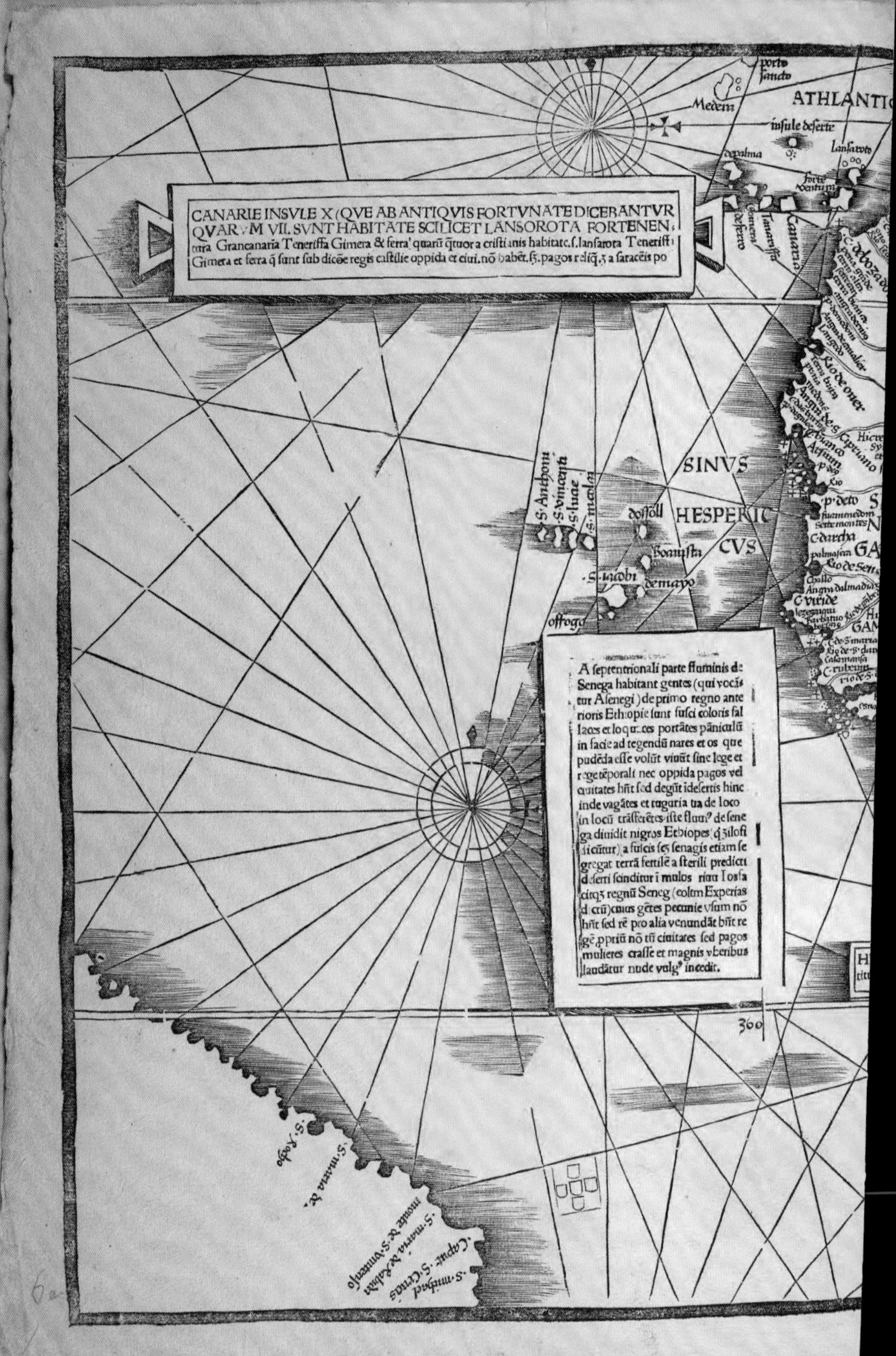
porto sancto
ATHLANTIC
Medeira
insule deserte
lansaroto
depalma
forte ventura
CANARIE INSVLE X (QVE AB ANTIQVIS FORTVNATE DICEBANTVR
QVARVM VII. SVNT HABITATE SCILICET LANSOROTA FORTENEN
tura Grancanaria Teneriffa Gimera & ferra' quarũ q̃tuor a cristianis habitatc. s. lansarota Teneriff
Gimera et ferra q̃ sunt sub dicõe regis castilie oppida et ciui. nõ habẽt. s. pagos reliq. 3 a saracẽis po
Gomera
Tanariffe
Canaria
SINVS
HESPERIC
CVS
S. Anthoni
S. vincenti
S. lucie
S. nicolai
dossoll
boavista
S. iacobi
demayo
offogo
A septentrionali parte fluminis de Senega habitant gentes (qui vocãtur Asenegi) de primo regno anterioris Ethiopie sunt fusci coloris fallaces et loquaces portãtes pãniculũ in facie ad tegendũ nares et os que pudẽda esse volũt viuũt sine lege et rege tẽporali nec oppida pagos vel ciuitates hñt sed degũt idesertis hinc inde vagãtes et tuguria sua de loco in locũ trãsferẽtes iste fluui⁹ de senega diuidit nigros Ethiopes (q̃ zilofi dicũtur) a fuscis ses senagis etiam segregat terrã fertilẽ a sterili predicti deserti scinditur i mulos riuu I osa citq3 regnũ Seneg (coltm Experias d cctũ) cuius gẽtes pecunie vsum nõ hñt sed rẽ pro alia venundãt hñt regẽ ppriũ nõ tñ ciuitates sed pagos mulieres crasse et magnis vberibus laudãtur nude vulg⁹ incedit.
360
C viride
S. Rocho
S. maria de
S. maria de Rabida
monte de S. Domenigo
Capus S. Crucis
S. michael

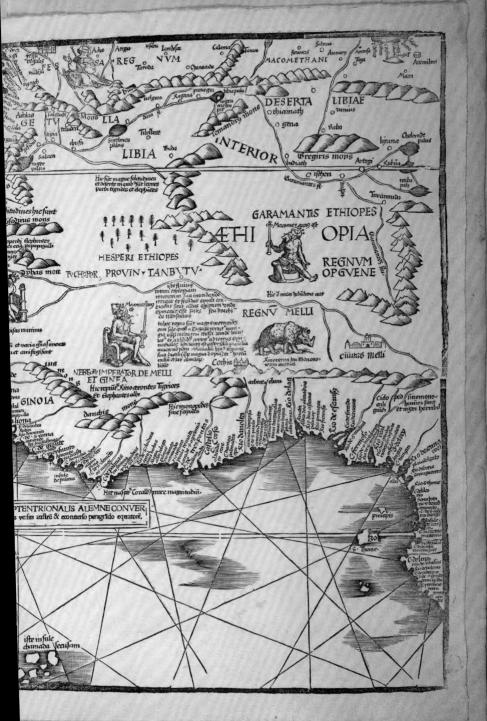

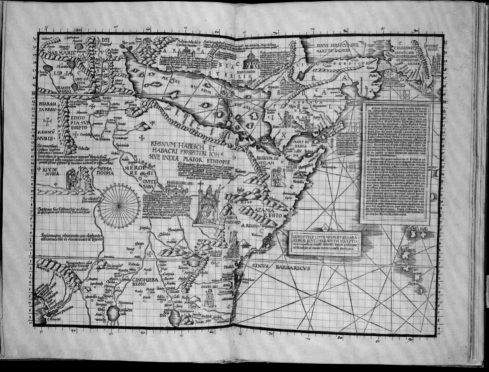

Figures 37 and 38
Sheets 7 and 8

Both of these sheets continue the general themes found in the other sheets of the *Carta marina* and populate the various geographical regions with characters like Prester John, who supposedly founded a Christian kingdom amongst the Muslims. Other elements of a cultural nature such as trade goods, precious stones, and spices are also depicted by Waldseemüller.

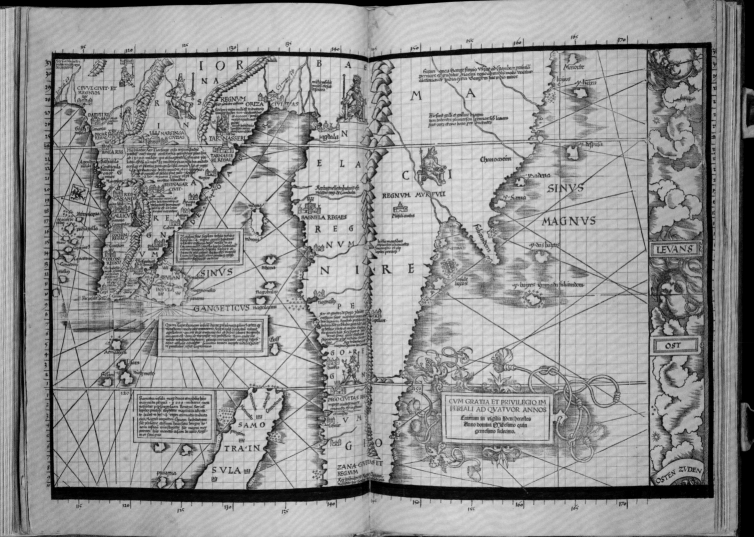

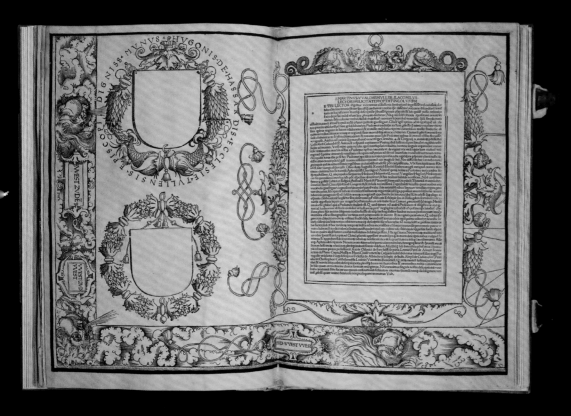

Figure 39
Sheet 9

This sheet contains the largest piece of text on the 1516 map. Schöner annotates this text by underlining particular areas in red and highlighting the name of Waldseemüller at the very beginning. One of the two blank shields on the sheet has a list of corrections printed underneath, all of which had been made to the map with the exception of those on printed sheet 6. This further indicates that the printed version of sheet 6 might have been a later addition to the *Sammelband*.

Figures 40 and 41 (page 90)
Sheets 10 and 11

Sheet 10 shows the southernmost extent of South America that Waldseemüller will show on the 1516 map. One is immediately struck by the difference of this representation with the full coastline on the 1507 map. On sheet 11 Waldseemüller will highlight what he terms "the newly discovered part of Africa." This is not different from that shown on the 1507 map and is referring to Ptolemy.

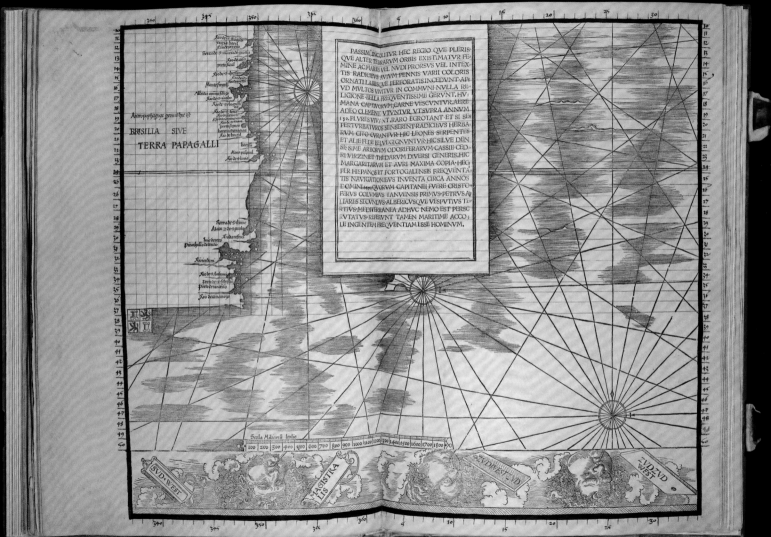

Figure 41

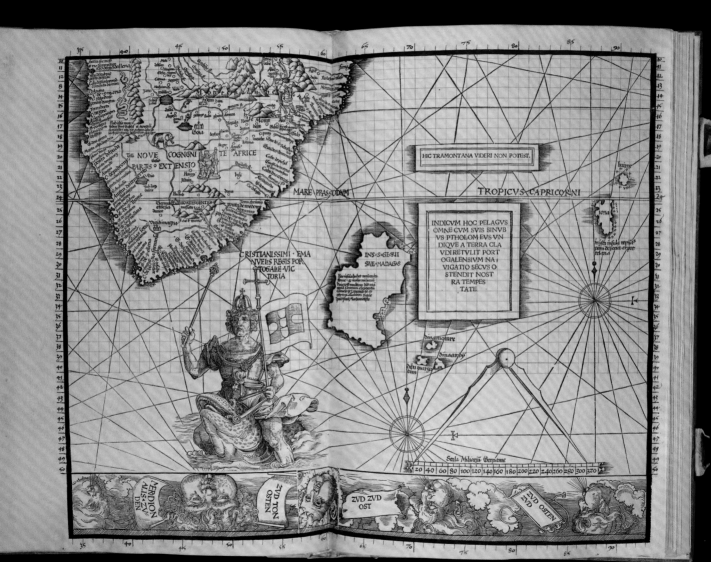

Figure 42
Sheet 12

This sheet is a true example of Waldseemüller updating himself and Ptolemy. Waldseemüller removes the large number of islands found on the 1507 map and in the *Geographia* from the 1516 map. One can also see that Schöner's grid is incomplete on the sheet as the lines of longitude do not continue into the non-cartographic border.

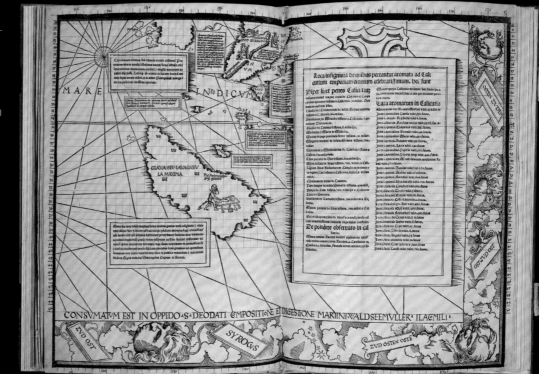

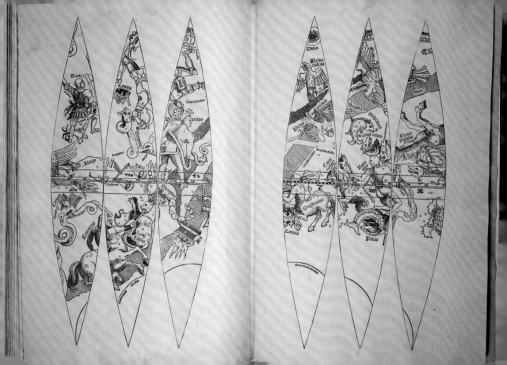

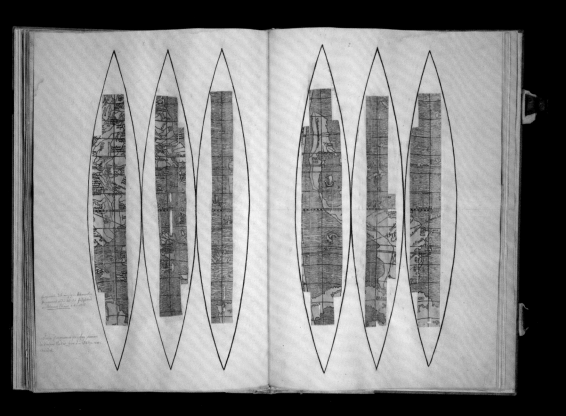

Figures 45 and 46
Terrestrial globe fragments

In 1903 the *Sammelband* was taken apart by Father Joseph Fischer in order to produce a facsimile edition of the 1507 and 1516 maps. In the binding, used as guards to attach the maps, were found both terrestrial and celestial gore fragments that are printed on vellum. These are the only known gores printed on vellum, which may indicate that they were a failed experiment by Schöner, as he was trying new methods for the construction of globes. The terrestrial fragments preserve some parts of Schöner's globe that indicate that he used the 1507 World Map of Waldseemüller as a source. One can see the separation in the Isthmus of Panama that is a signature of the Waldseemüller map, along with the name America printed across the southern continent. Besides these features Schöner goes beyond Waldseemüller's radical vision of New World geography by including the full sea passage around South America, something unknown in 1515, the date of the gores.

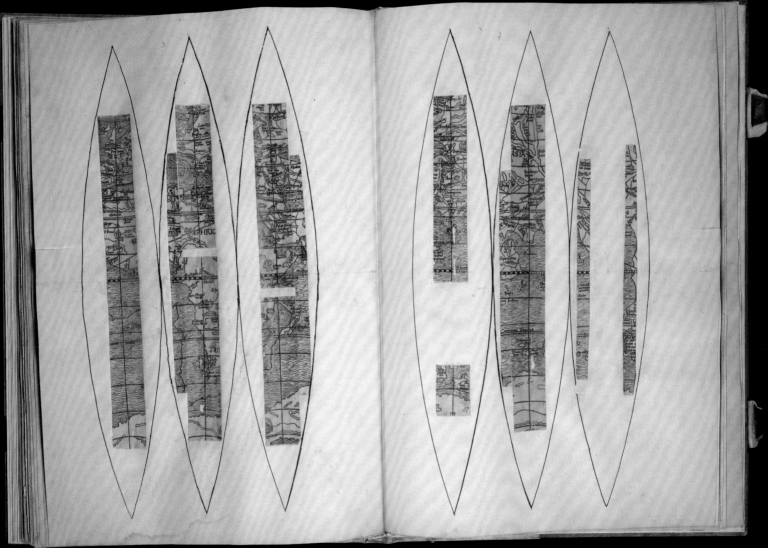

Figure 47
Celestial gore fragments

The Celestial gore fragments that were found in the binding of the *Sammelband* are different from the full gores and show a higher state of completion. The fragments display both the lines for the Tropics and the ecliptic but otherwise are identical to the full gores.

Figure 48
Dürer–Stabius star chart
Courtesy of the Collections of Wolfegg Castle

The star chart made by Albrecht Dürer and Johannes Stabius, which Schöner bound into the *Sammelband*, is the earliest printed star chart known. The charts that Dürer produced typically came in pairs of the Northern and Southern hemisphere. The only chart that remains in the *Sammelband* was that of the Southern hemisphere and it is heavily annotated by Schöner.

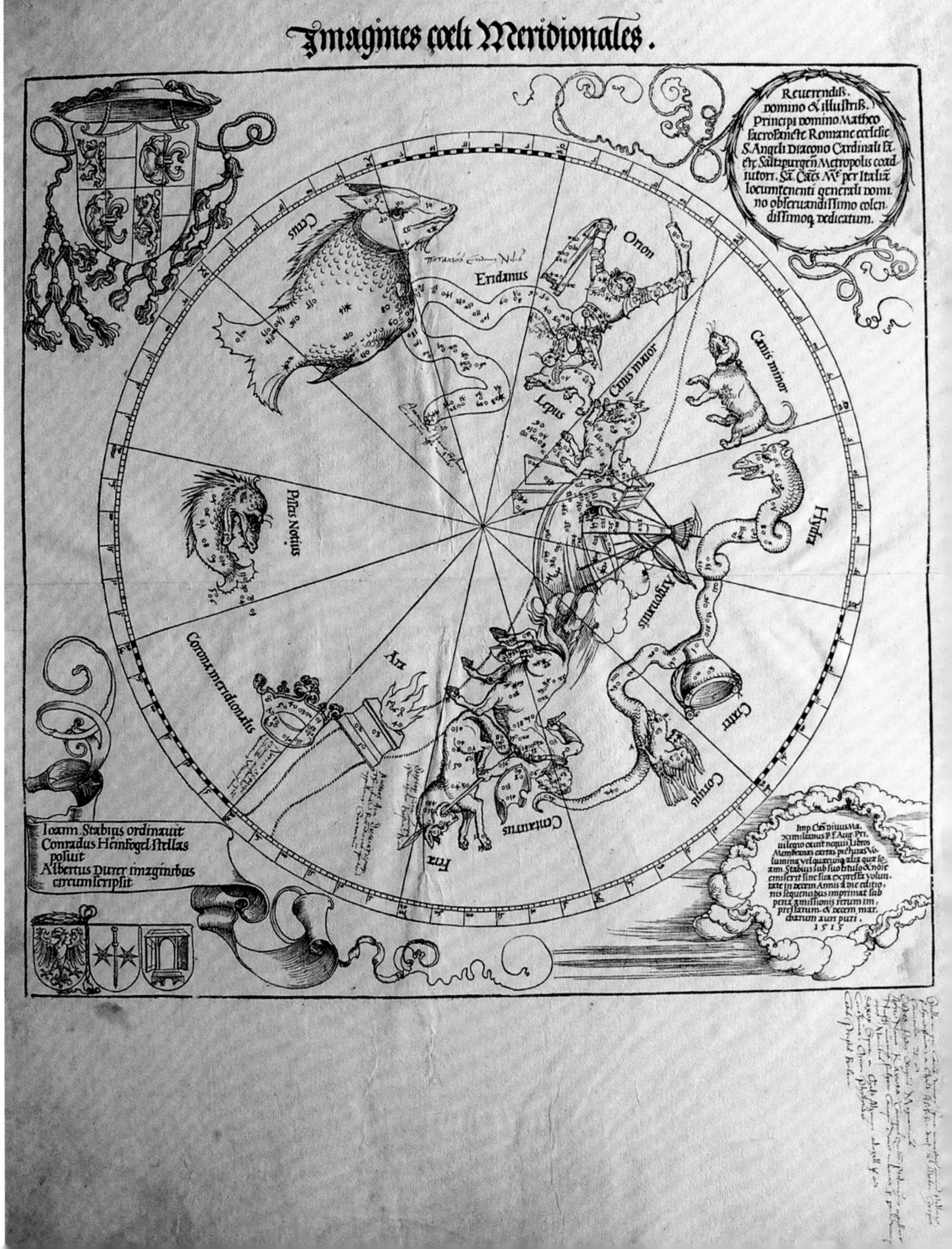

e triũ & ſemis cũ tertia taliũ
ẽta cẽtũ & octuagīta & vni
tertia.& G punctũ ad quem
ies in tabula plana paralelli
ponatur A B C D.tabula i
iterum habens.A B.q̃ A C.
e ad ipas directã habeat E F
qualis e.E F·recte ad nonagi
um partẽ.Dimiſſis igitur i F
tia & duodecia.& i G H ui
ẽ ſemis cũ tertia.& G K eo
s ſexaginta & tribus.Et poſi
tiali.erit q̃de H·circulus p ſy
edio habitabilis terre locat⁹
lus erit:auſtrale plaga habi
s:& oppoſitus circulo p me
irculus erit·p quẽ terminabi
trionalis:p inſula tyle ductuſ
a linea:earũde ſectionũ cẽtũ
uni⁹.& ſemis cũ tertia aut ſo
ctuaginta cũ in nulla memora
deſignatio a ſe differt a cẽ
tias F & H & K deſcribem⁹
tq̃.O & H & P atq̃.M & G
etias.Propria igitur ratio pa
nationis ad planũ axis ipius
ſeruabitur cũ & hic axis incli
H & rectus eſſe ad planum ta
ter oppoſiti fines deſignatõ
cõprehendantur·Vt autẽ &
qlis latitudini ſit·cũ in ſpera
⁹ circulus q̃nq̃, taliũ fere pa
eſt·duo cum q̃rta p ſyene au
nis cum duodecia·p merõem
s cum tertia opuſq̃ e·ad vtrã
ridionalis·F K.dece & octo
ianos.p tertia ſcilicet pte vni
ctialis vt cõtineatur p ſemicir
gitudinis ſpacium:ſumemus
n quelibet trium paralleloꝛ q̃
q̃ gradib⁹ p tertia pte vnius
de p duos gradus cũ q̃rta ſe
s qliũ.E F·e nonagita·Ab
tuor & ſemis cum duodecia

Ab F autẽ cum quatuor & ſemis cum ter
tia in eiſde ipis.Poſt hec ſcribetes p tria pũ
cta diſtantiaꝛ equipollentium circuſerẽtias
q̃ erũt p reliq̃s meridianis velut tminantes
totam longitudinem ſcilicet.S T V & X y
Z.ſupplebimus circuſerẽtias p ceteris pa
ralellis a cẽtro q̃de L p ptes vero notatas i
F K.ſcdm diſtatias ipaꝛ ad ipm eqnoctia
le.Qd aũt modus hic magis ſit ſimilis ſpe
re q̃ alius hinc clare patet·Cum illic manete
ſpera nec circũducta q̃d tabule cõtingit ne
ceſſe e cũ aſpectus in medio deſignatiõis fi
gitur vt vnus q̃de meridianus:q̃ medius e
& ſub axe noſtri aſpectus cadit.imagine re
ctã linee p̃beat.Reliq̃ vero q̃ ex vtraq̃ pte
iſtius ſint·omes vertantur ad hũc i ipoꝛ
curuatoibꝰ & magis q̃ ab ipo plus diſtet
q̃d hic aiaduerteretur cũ decẽ curuationũ
pportione·Preterea coeqtione circuſerentia
rũ paralelloꝛ inuice nõ ſolũ ad eqnoctiale
& ad paralellũ p tyle queadmodũ illic e p
pria ratione habere·ſed etia in alijs q̃ maxi
me poſſibile ſit veluti intueri ſas e.Inde to
tius etia latitudinis ad totã lõgitudinẽ neq̃
ſolũ in paralello p rhodũ vt ibi ſed fere in
omibus·Si eni hic pducamus.S & V recta
queadmodũ in priori pictura.H.& circuſe
rentia minorẽ ratione habebit ad F S & K
V·quã oportet i p̃ſenti ſigura cũ cõprehẽ
ſa hic ſit p totã H T·Qd in eqnoctiali p̃ter
accidit G M.Si vero coeq̃lem hanc faciem⁹
ad K F·latitudinis ſpaciũ cũ F S & K V·
maiores erũt q̃ coeq̃tiones ad F K·veluti
K T·Si aũt F S·& K V·ſeruemus coeq̃les
H.& mior erit ad K F·q̃ coeq̃tio veluti H
T·Ex bis igitur mod⁹ iſte melior habetur·
q̃ pmus ſed ab illo etia deficiet in facilitate
deſignatõis·cũ illic ab vnius ꝛegule circum
ductõe:deſcripto vno paralello diuiſoq̃, lo
cari poſſit qlibet locus·Hic aũt nõ ſimiliter
cõtingit ob meridionaliũ lineas ad media
flexas.Omes eni circulos inſcribere ſingilla
tim oportebit & locoꝛ ſitus inter palellos
incidentes ex vtrorumq̃, ratiõibꝰ coniectare

His aũt ſic habitis magis & hic tenendũ e·
qd ſit equius q̃dq̃, ſerioſius·q̃ id q̃d debiliⁱ
faciliuſq̃·Vtreq̃, tamẽ forme ſeruãde ſunt.
ob ea q̃ facilius in opere adducuntur.
Qualiũ e eqnoctialis q̃nq̃, taliũ e p merõem
q̃tuor & ſemis cũ tertia.Vnde ratione ha
bet ad ipm q̃ triginta ad viginti & none·
Qualiũ e eqnoctialis q̃nq̃, taliũ e p ſyene q̃
tuor & ſemis cũ duodecia·Vnde rationem
habet ad eũ q̃ ſexagita ad q̃nq̃ginta & q̃n
q̃ boc e q̃ duodeci ad vndecim·
Qualiũ e eqnoctialis q̃nq̃, taliũ e p rhodũ
q̃tuor.Vnde rationẽ bet ad ipm epitetarti·
Qualiũ e eqnoctialis q̃nq̃, taliũ p tyle duo
cũ quarta.Vnde ratione babet ad ipſum q̃
viginti ad nouem·

Claudij ptolomei coſmographie liber pri
mus explicit.
CLAVDII PTOLOMEI VIRI AL
LEXANDRINI COSMOGRAPHIE
LIBER·SECVNDVS·HEC HABET·
Eiuſdẽ tractatus expoſitione plage magis
occidẽtalis Europe iuxta has puintias ſeu
ſatrapas.Britania.Iſpania.Gallia·Germaniã
Retiã·Vindelicos.Noricũ·Pannoniã· Illiri
cos.atq̃ Dalmatiam.

CHAPTER 1

CARTOGRAPHY IN THE MARGINS

How Johannes Schöner Read His Maps

Figure 50 (detail)

One of the most remarkable features of Schöner's extant manuscripts comes from his annotations in the various editions of Ptolemy's *Geographia* that he owned. This group of atlases, which is today owned by the Österreichische Nationalbibliothek, is characterized by a great number of handwritten corrections and complex annotations that show his thinking about theoretical cartography and the state of the art in the early sixteenth century. Schöner's library contained some of the most important books on cartography and geography that were available to him at the time, including copies of the 1482, 1509, and 1513 editions of Ptolemy's *Geographia*, and how he went about reading them and using some of his other maps can tell us a great deal about how the science of cartography was perceived during this critical period in its development.[23]

The importance of the rediscovery of the text of Ptolemy's *Geographia* to the study of cartography and geography of the late fifteenth and early sixteenth centuries would be difficult to overestimate. The text was part of the recovery of Greek science and Ptolemy's 'handbook' united astronomy, mathematics, and geography into a single systematic method for describing the known world graphically, in the form of maps. The transmission of the text of the *Geographia* is an extremely problematic issue in the study of early geography and cartography. At present there survive fifty-three Greek manuscripts of the book and none of them were written earlier than the thirteenth century.[24] This presents scholars with at least a one-thousand year gap between the writing of the original text and the earliest evidence we have for it.

In most of Italy and Germany, during Schöner's lifetime, the text was not known from Greek manuscripts but only from a flawed Latin translation made by Jacopo Angeli da Scarperia at the beginning of the fifteenth century. The Greek language was not well known during the early Renaissance and so the translation of Ptolemy's text into Latin made it more widely available to scholars and humanists interested in the recovery of the geographical sciences brought on by the new explorations around the globe. The first major revival of the study of Greek took place in Florence in the late 1300s and it is here that interest in the recovery of Ptolemy's text first appeared. It was in Florence, from about 1397 to 1399, that Emmanuel Chrysoloras began to teach Greek to a small group of young and brilliant scholars who would later make some of the more important translations of Greek texts into Latin, such as those of Ptolemy.[25]

Of all the newly discovered texts coming from the East, it seems certain that the text of the *Geographia* piqued the interest of many of the new Greek scholars. Florence during the late fourteenth century was a city of international trade and shipping, with many of the most important merchants interested in understanding the geographical sciences and in picturing the extent of the known world. One of the members of Chrysoloras' classes, the trader Palla Strozzi, purchased a Greek manuscript of the *Geographia* while in Constantinople and brought it back for the group to translate. Even though this was probably the first introduction of the *Geographia* back into Western Europe, fragments and quotes that alluded to its content survived in earlier Latin texts.

Mention had been made of Ptolemy from the sixth century onward, first in works by the late-antique historian Jordanes, and later in astronomical treatises translated from Arabic. Although the extracts of the book appeared in many different forms, it was probably not possible, at least from what was then known, to reconstruct the actual method for drawing a flat map of a sphere that Ptolemy laid out in the *Geographia*.

The hunt for Greek manuscripts of Ptolemy dominated many of the humanist book hunters of the period and finding them was a prize many wanted. One of the best-known of the book hunters, Poggio Bracciolini (1380–1459), wrote in a letter to his friend and fellow bibliophile Nicolaus De Niccolis that he hoped to live long enough to capture it, for "[he] should like to see any sheet from Ptolemy's *Geographia* if it can be done; remember this in case you should chance by a page."[26]

The transmission of the text of Ptolemy to the West through early sources is still a problematic area for scholarly research as there are tantalizing hints in the writings of humanists like Jean Fusoris. That he knew the text of the *Geographia* by the end of the fourteenth century is apparent from the end of his commentary on the *Sphere of Sacrobosco*, where he writes:

> And this technique amongst others the wise Ptolemy used in compiling a table of longitudes and latitudes of cities, and in making his book of the mappemonde which is in the library of Notre Dame de Reims, and also in making the book of the marine chart,

because he sent several learned astrologers east and to different cities, who, through the above-mentioned technique took first the longitude of the cities; that is, which city was more to the east or to the west, and by how much and to what degree. And also by using the astrolabe and other instruments of astrology, they took the elevations of the North Star or of another fixed star, and so they had the latitudes of the cities too … and thus they had their exact situation on the surface of the earth.[27]

Fusoris' description fits the general pattern of the *Geographia*, being that it is for the most part a listing of latitudes and longitudes of cities and towns throughout the known world.

The translation that brought Ptolemy into the hands and minds of western scholars was accomplished by Jacopo Angeli da Scarperia. In the preface to his translation, which would be used in most of the early printed editions, Jacopo says that the reason a translation of the *Geographia* is needed stems from the difference in perspective found in the studies of geography produced by Latin and Greek classical writers. Latin geographers, according to Jacopo, are un-mathematical, his prime representative of the Roman approach being the *Natural History* of Pliny the Elder. Greek writers, on the other hand, are concerned with the problems of depicting scale, showing longitude and latitude, and measurement. Jacopo writes that "none of our Latin writers explains how our globe, which is spherical, can be described on a flat surface," and it is for this reason that Ptolemy

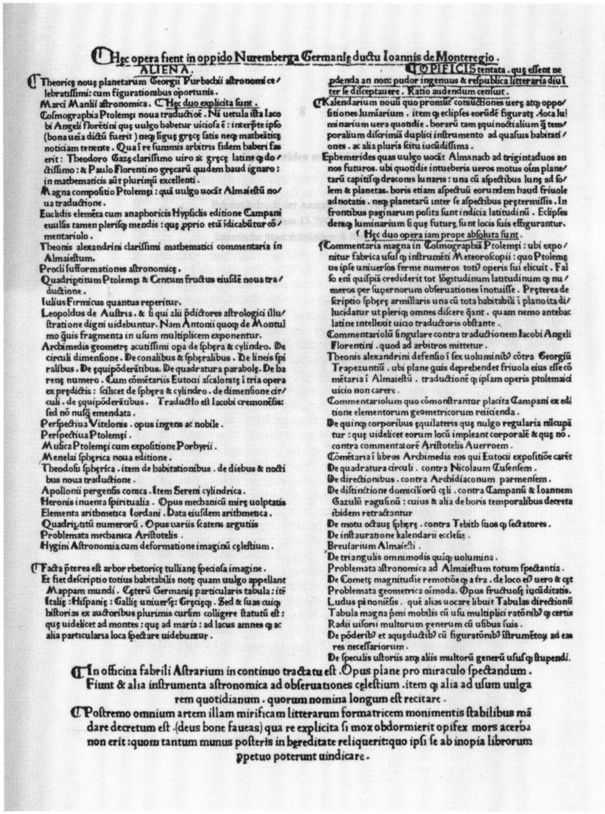

Hęc opera fient in oppido Nuremberga Germanię ductu Ioannis de Monteregio.

In officina fabrili Astrarium in continuo tractatu est. Opus plane pro miraculo spectandum. Fiunt & alia instrumenta astronomica ad obseruationes cęlestium. item & alia ad usum uulgarem quotidianum. quorum nomina longum est recitare.

Postremo omnium artem illam mirificam litterarum formatricem monimentis stabilibus mādare decretum est. (deus bone faueas) qua re explicita si mox obdormierit opifex mors acerba non erit: quom tantum munus posteris in hereditate reliquerit: quo ipsi se ab inopia librorum ꝑpetuo poterunt uindicare.

Figure 49

Regiomontanus' Index

The list shows the ambitiousness of the publishing program that Regiomontanus sought to undertake before his death in 1476. Many of the great classical writers on mathematics like Euclid, Ptolemy, and Diophantus appear as part of his program. Regiomontanus would get to only a fraction of what he hoped print, with Johannes Schöner publishing only a fraction more by 1547.

needed to be translated.[28] Jacopo was, however, no great translator of Greek and by the mid-fifteenth century many readers began noticing problems with some of the more technical parts, especially the astronomer Regiomontanus, whose manuscript corrections to the *Geographia* would come into the hands of Johannes Schöner.

Schöner was an active editor of the mathematical and scientific works of Regiomontanus (1436–1476) and, as has been mentioned earlier, published many of the works that Regiomontanus had begun, but was unable to complete. Regiomontanus, in a small print shop he set up in Nuremberg, published an index of works that he wanted to edit, collate, and correct. On the list is an edition of Ptolemy's *Geographia*, which comes directly after Peurbach's *Theorice novem planetarum* and Manilius' *Astronomica*. Although he never published his edition of Ptolemy, the manuscript that he left behind came into the hands of Johannes Schöner and is now in the Universitätsbibliothek in Basel. The first page reads, "The Geography of Ptolemy—in composition and numbers and to the Greek (and Latin added in the margins) editions—edited by Jo Regiomontanus." This first part of the manuscript title is in the hand of Regiomontanus; the title continues in the hand of Schöner, where he writes that he purchased it in Nuremberg from Willibald Pirkheimer (1470–1530).[29]

During his life Regiomontanus was among the first readers of Ptolemy's geographical works to recognize that the Latin translation currently available was far from accurate. In his *Dialogus adversus Gerardum Creminensem in planetarum theoricas deliramenta*, Regiomontanus expresses his concern with the Latin translation of the *Geographia* by asking,

> What will happen if the first copy has been rendered obscure by a careless translator, or transformed by the first starving copyist who happens along? Both of these things can be seen in the work that today is passed off as being Ptolemy's *Geographia*, in which the literal structure intended by the Greek author does not correspond to the phrases written by Jacobus Angelus … who mistakes the meaning of words and in which the appearance of the maps of the specific provinces do not preserve the appearance intended by Ptolemy.[30]

Regiomontanus continues by critiquing the mathematics and technical knowledge of Jacopo and concludes,

> As a result a person who thinks he has Ptolemy's *Cosmographia* [*Geographia*] at his disposal could not even bring forward the palest shadow of that great work; and, without exception, the entire world will believe me when I say that, in effect, this work has not yet been handed down to the Latins.

Schöner, in his reading of Ptolemy's geographical and mathematical texts, displays this same concern with the problem of accuracy and the reliability of the original sources. In the preface to his edition of Regiomontanus' *On Triangles of Every Sort*, printed in 1533, Schöner emphasizes the care he takes in editing and using sources, saying "Our task has been to transcribe with the greatest fidelity everything from the original text."[31] All of the editions of Ptolemy that remain extant in his library reflect this, displaying a great many numerical, technical, and methodological corrections as marginal annotations.

While we cannot be certain which of Schöner's Ptolemaic atlases he may have annotated first, we are sure that his copy of the 1482 Ulm Ptolemy came into his possession in 1507. According to an annotation in the text that is in Schöner's hand, he purchased the book on October 16th of that year.[32] The book, which is bound between heavy wooden covers connected with leather-backs that show blind imprinting, all in the same manner as the *Sammelband*, also displays Schöner's bookplate and contains a number of manuscripts in Schöner's hand. The Ulm edition of Ptolemy is one of the great masterpieces of early atlas printing and its world map is an iconic symbol of fifteenth-century cartographic art.

The manuscripts found in Schöner's Ulm Ptolemy[33] are *De locis ac mirabilibus mundi et primo de tribus orbis partibus*, together with the *Registrum super tractatum de tribus partibus*, the *Registrum alphabeticum super octo libros Ptolomei*, and the *De mutatione nominorum*. Schöner's copy of the 1482 *Geographia* is heavily annotated in his hand throughout the book and also on all of the maps that it contains.

The majority of Schöner's annotations are of a highly technical nature, and concern specific corrections to latitudes and longitudes as well as

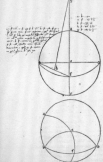

duo mẽ trecentī ac sexagíta. Relíqua vero
angulus sub HGF vigintí ac nouẽ cum ter
tia q̃ dúo mẽ trecentí ac sexagíta. Relíqio
vero angulus sub H·GF·vigíntí & noue cũ
tertia obq̃, id ratio G·F·erat F·H·Eadéq̃ &
cntía ac octaugíta ac vnius ac sexíes cũ ter
tia ad q̃dragíta ac fiex ac fexies cũ vicesima.
Et q̃ síqz H·F·ésta q̃dragíta & fex ac fe
mis cũ vícesimétalis ẽ·B·E·ésta nonagíta
Vnde cũ q̃fí ẽ E·B·ésta nonagíta sit FE·

Hís síbata ponátur A·B·C·D·tabula i
duplo maíorí íterum habens A·B·q̃ A·C·
effi aút A·E·ac sid iusto directē habet E·F·
dínidarurq̃, equalis E·F·ésta ad nonagíta
te quarta ídeuem parte. Diuisíte igitur IS·
G·deuís cũ tertía ac duo deciá. ac í GH·vi
gínti ac tríbus ac femis cũ tertía ac GK·no
rudem ficisbus ficsqsíate ac tríbus. Et posí
to·G·p· equoctíalí. erit q̃d H·circulus p̃ fy
esat ferme ín medío síbdatíon ínter loca?
F·vero parallelus erit ad hunc plaga habí
ta tíonis tunc uíce oppositus círculo p̃ me
nem·K·cí círculos erít q̃ erat terminaliae
ha líne ficptenrionalíg ín liac tyle dúbuf
deinde p̃ ducta línea vicesíe ficstíoní oms ac
ac octaugíta vní· ac femís cũ tertía tot fo
lum cents ní octaugíta cí ín iulte mentos
bili vnís ob íd defígnatíu ẽ fer differt a ce
to·L· distantías F·ac H·ac K· defcríbem̃
Qu·K·ac H·oq̃, O·ac H·q̃ regq̃. M·ac G·
q̃ N·círcueferẽtías. Propria igitur mõ p̃
ralellos· inclinationis ad plana vníe ípsua
aspectus síc obferuabitur sic ac hic vnío ínclí
nant defigl· H·ac tellus síle ad planum ta
bule et equalíter oppositi línec defígnatu
nís ín aspectu cõprehendantur·Vt aút ac
longínudo crescíte latitudíní líeerit ín íper
glín· q̃ maíor crealus q̃q̃, tális sere p̃a
rálellus p̃ tysil·e st·duo cum íprt p̃ fysiea ac
síc q̃ ísnor ac femís cum duoden· p̃ meream
q̃nor ac femís cum tertía obsq̃, d ad vníe
q̃ p̃rí línet merídíanus p̃·F·K·decú ac octo
scríbem merídíanus p̃ tersíos sicíst p̃·vní
ac fiexe equocíalis vt cõnecntur q̃ líeeue
cales ersí longítudíne spacíum dínenstus
festíones sídem q̃líbet tríum parallelos q̃
eq̃iualet· q̃q̃, q̃disce p̃ tertía p̃ vníus
hor ab R·qualis p̃ duos gradus cí q̃rt fe
dínous facíete q̃luíon·E·F·e nonagíta· Ab
H·vsto p̃ quatuor ac semís cum duodecía

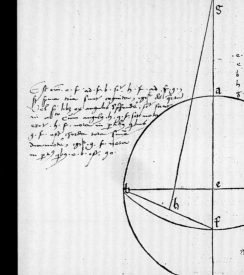

Ab·F·aute cũq̃ainqz ac femís ite tertí
tís ín eífde ípfis·Post haec fimbae p̃ tria p̃
iffa distantíae ac qoestíonem ac críferítíae
q̃ erít p̃ víliq̃mochanís velut límínantas
totam longítudínem líber·S·T·V·x·X·Y
Z·fupplétíous círcũferentías p̃ ceterís p̃a
ralellía a céto q̃d·L·p̃ p̃tes vero notabís í
F·K·óbem distántías q̃sẽ ad ípsm equocíalẽ
líc·Q·si aút modus hic magís sit fimíle ípes
se q̃ alíus líene cũm patet·Cum íle magís
ípsea aec círculados q̃ tabule cõfigat q̃ ma
críle ã cí aspectue ín medío defígnatíc g̃
gítur vt vnus q̃dẽ merídíanus q̃ medíuas
ca síb aee retíla líea dúceret ín alíos q̃ ín
tís líene plínate·Releq̃ vero q̃ ex vtraq̃ p̃te
ístua íter· omes vertuntur ad hoc í ípoã
curuatíoni ac magís q̃ aby ípo plus díffert
q̃d híc aúsdoctríne cí decus curuatíons
p̃portíoe·Praeter q̃ositíone círcũferentía
rũ parallelor ínuicé nõ solú ad equocíalẽ
p̃d ad paralelly p̃ tyfí· q̃íadmodu illíc p̃ p̃
príe ratíoní líber·fi·q̃dẽ vt ín tertí magís
nõe possíbíle síc· velut íí íntuerí fíes·Inde tu
tíus etía latítudínis ad tríS ĩ q̃ sídínabíe noq̃
solú ín parallelo p̃ rhodís vt síc sed fere ín
ómíbus·Síc melíor·p̃ducerem. S·a·V·vtdí
q̃íadmodu ín príorí plínua·Hac círcũfere
uentía ínuení ínclínanhabítet·ad·F·K·veluí
accídit G·M·Si vero q̃sẽh haec facíem̃
ad·K·F·latítudíne líeetdi·ín·F·q̃·e·K·V·
maíores erít q̃ ea q̃nõesmes·ad·F·K·veluí
K·T·SiaúE·F·ee·K·V·íeeuenus q̃sẽ íp fa
Hac níee erit·ad·K·F·q̃ q̃alía velut·H·
T·Ex hís ígítur mõ q̃ úthe melíor habeue
í∫ genus ficd ab illo mã defícít ín facílíte
delígnatíoís ílle vt· vnius regule círcã p̃
ductõe defígto vto paralelloş deslíg̃ fe
rarí posset·Quore latíor locus·Híc aút sõ símílíter
cõfugíot ob merídíanus líe∫ at medíã
fexoa·Oñes ení círcuíos ínferíore síngulís
títe oportebít ac lõq̃, líme íter paralelos
ín∫dentís ex vtrorŭ, taríoís vemelĩ

Claudii ptolomei víri alex̃drini cosmographie líber fri
mus explicit.

CLAVDII·PTOLOMEI·VIRI·AL
EXANDRINI·COSMOGRAPHIE
LIBER·SECVNDVS·HEC·HABET·
Eiusdẽ tractatus expositiões plage mappe
occidentís Eurpe iuxta hũc p̃otatas líte
sempus Britaní·Hispaní·Gallí·Germaní·
Retí·Vindelícia·Norící·Pannonía·Illeri
os.ac.q̃.Dalmatíam.

the three map projections that Ptolemy discusses and illustrates in the text. A map projection can be thought of as a way of keeping track of the error induced into a map by the mere process of taking the surface of the globe and flattening it out. If one thinks of the skin of an orange, for example, one cannot remove it from the orange and flatten it without either tearing it or distorting it in some way. The same can be said for the surface of the earth, as compromises in the accuracy of distances, area or shape must be made in order to make that surface into a flat map. It was Ptolemy in the second century who first figured out a mathematical and geometrical method for consistently keeping track of this error and how it might affect the look and accuracy of a map.

Many of Schöner's annotations on Ptolemy's projections simply repeat some of the numbers found in Ptolemy's text and re-state the distance between points on the figure and some important ratios. Other annotations and projection studies are more detailed. Immediately after the manuscripts mentioned above there are three drawings glued into the atlas on sheets of paper that are much smaller than the other leaves that make up the atlas.

Here, rather than simply reading the text we see Schöner actually trying to understand in detail the construction of the projections, which for a newcomer to the *Geographia* is not an easy thing to do. The so-called First Projection of Ptolemy, which Schöner draws, is the simplest of the three projections and the first to be discussed in the *Geographia*. During his presentation of the geometric methods needed to

produce a map on the projection, Ptolemy also inserts remarks about its effects on the viewer.[34] For example he begins by stating the geometric criterion,

> It would be well to keep the lines representing the meridians straight, but [to have] those that represent the parallels as circular segments described about one and the same center, from which (imagined as the north pole) one will have to draw the meridian lines.

After describing the geometrical part Ptolemy states the reason for the construction,

> Above all, the semblance of the spherical surface will be retained … with the meridians still remaining untilted with respect to the parallels [i.e., perpendicular to them] and still intersecting at that common pole.

In the case of the first projection that Schöner drew in the back of the 1482 Ulm *Geographia* there are three such pairs of geometric criteria describing why the mathematics is being employed and how it is changing the distances and measurements compared with those on the surface of a sphere.[35] The second criterion states that "since it is impossible to preserve for all the parallels their proportionality on the sphere, it would be adequate to keep this [proportionality] for the parallel through Thule and the equator." Schöner's annotations seem to note the problematic nature of this proportionality and show him attempting to calculate what is happening to

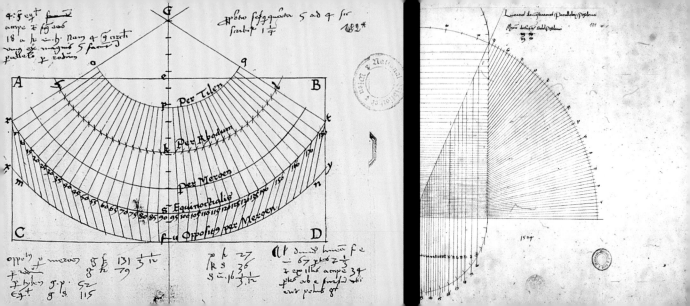

the scale of the map on some of the other parallels. The effect of Ptolemy's method is that the east–west distances are proportional to the north–south distances only along the parallel through Rhodes, and are distorted more progressively the farther one moves away from that line.

The third, and perhaps most interesting, of the three drawings found in Schöner's copy is somewhat mysterious and not easily understood. This diagram, entitled *Lineares demonstrationes Parallelorum Ptholemei,* is a type of computational device that allows the continuous conversion of the length of the longest day on most parts of the globe to the latitude of that location and the corresponding parallel of Ptolemy. The problem of understanding Ptolemy's conception of latitude and the operation and uniqueness of this calculator is a case study in the type of geographical and mathematical work on which Schöner labored his entire life.

The concept of latitude in Ptolemaic astronomical and cartographic theory is complex and is quite different from the modern notion of a group of equally spaced parallel lines on the surface of the globe or a map. For Ptolemy, latitude was principally an astronomical concept, an angle of inclination, which varied with the location of an observer and determined which stars were capable of being seen in that location. In Ptolemy's writings, especially in the *Almagest,* he assumes that the observer is at intermediate latitude, somewhere in the northern hemisphere, and therefore, that the stars in the possible universe fall into three groups. The groupings are based on observability and include the stars that

never set but are always above the horizon, the stars that both rise and set, and the stars that never rise in that location and are therefore always invisible. Using two parallel lines of equal size, Ptolemy separates these three groups of stars on the celestial sphere.

The two circles used to separate the groups of stars were also used by Ptolemy, and all classical geographers, to define what we now know as compass directions. For example, as one proceeds northward from the equator the circle of always-visible stars will be seen to increase until one reaches the North Pole at which time it will coincide with the line of the horizon, while at the same time the circle of invisible stars also increases. Ptolemy demonstrated that a locality X is north of some locality Y just in the cases where some star that is always visible at X, rises and sets at Y, or if some star that cannot be viewed at X, rises and sets in Y. Because of the fact that all these phenomena were seen not to change if we move from east to west on the earth's surface, they were used to define a parallel of latitude. Hence latitude is in general defined astronomically, rather than terrestrially for Ptolemy.

In Book II of the *Almagest*[36] Ptolemy explains that there are many types of phenomena that are characteristic of latitude:

> The individual points [concerning the *sphaera obliqua*] which might be considered most appropriate to study for the subject we have undertaken are the more important phenomena which are particular to each of the northern parallels to the equator and to the region of the earth directly beneath each. These are

1. the distance of the poles of the first motion from the horizon, or the distance from the zenith from the equator, measured along the meridian;[37]
2. for those regions where the sun reaches the zenith, when and how this often occurs;
3. the ratios of the equinoctial and solstitial noon shadows to the *gnomon*;[38]
4. the size of the difference of the longest and shortest day from the equinoctial day and all other phenomena which are studied concerning;
5. the individual increases and decreases in length of day and night;
6. the arcs of the equator which rise or set with arcs of the ecliptic;
7. and the particulars and quantities of angles between the more important great circles.

In section 6 of Book II Ptolemy describes particular characteristics of the various parallels and defines their exact locations on the surface of the earth using constant increments of the longest day at various locations. The latitudes corresponding to this regular series of daylight increments are not equally spaced but become more crowded the farther one moves from the equator. This way of defining latitude on a map produces a form of graticule, or coordinate grid, that differs greatly from that found on modern maps and it is the attempt to understand this relationship that caused Schöner to produce the

diagram.[39] Schöner's sketch reproduces Ptolemy's relationship between length of day and latitude in a unique geometrical way that allows one to quickly convert from one to the other.

Ptolemy reproduces the values of the length of day and latitude that he gave in the *Almagest* in Book I chapter 23 of the *Geographia* as follows (I do not produce the entire list, just some examples for use in discussing Schöner's calculator):

These limiting meridians will enclose twelve hour intervals according to what has been demonstrated above. However we have decided it is appropriate to draw the meridians at intervals of a third of an equinoctial hour, that is, at intervals of five of the chosen units of the equator, and to draw the parallels north of the equator as follows:[40]

1. The first parallel differing by ¼ hour, and distant from the equator by 4¼ degrees, as established approximately by geometrical demonstrations.
2. The second, differing by ½ hour, and distant 8⁵⁄₁₂ degrees.
3. The third, differing by ¾ hour, and distant 12½ degrees.

…

10. The tenth, differing by 2½ hours, and distant 36 degrees, which is drawn through Rhodes.

…

20. The twentieth, differing by 7 hours, and distant 61 degrees.

21. The twenty-first, differing by 8 hours, and distant 63 degrees, is drawn through Thule.

Schöner's calculator reproduces this list graphically, allowing for the conversion of length of day to latitude and of latitude to length of day in a way that would have been very convenient for a map- and globemaker of the early sixteenth century. These kinds of graphical aids are quite common throughout Schöner's writings and we shall see several in the course of the book.

To get a sense of the graphical operation of the sketch we need to first look at the bottom of the figure, which shows a curved line marked with numbers from 12 to 20 and that continues unmarked for three more divisions, making a total of 24 intervals. These divisions represent the hours in a day past the 12-hour period that makes up the equatorial day. In Ptolemy's list of hours that define map parallels, he expresses the duration of the length of the longest day at various latitudes as an additive difference from this 12-hour day. Schöner displays the difference geometrically, allowing for the relationship of the latitude to the time of the longest day to be calculated in a continuous and not simply discrete way. Above the curved lower line in the figure there is a series of additional numbers marked from 0 through 21 representing the numerical sequence of parallels defined by Ptolemy in the above list. For example, if we observe the number 21 we can see that it corresponds to the number 20 on the curved line below it. The parallel 21 of Ptolemy is the line that runs through Thule and has a difference of 8 hours in the duration of its longest

day from the equatorial day. In this case, because Schöner's beginning point on the curved line is 12 hours, representing the equatorial day, the addition of 8 hours gives the number 20 shown on the curved line. To use the computer to calculate the number of degrees corresponding to any particular length of the longest day, one follows the vertical line on the figure, continuing with the 20-hour line as an example, until it intersects the diagonal line that bisects the center of the drawing. One can then follow the horizontal line at the intersection over to the right to determine the latitude. In the case of the 20-hour line we find, using Schöner's diagram, the latitude of 63 degrees, corresponding with the values that Ptolemy lists.

The construction of the calculator is more complicated than its use. The curvature of the lower line is based on the variation in great circle distances of latitudes and the change in the length of day. One can see this variation by simply looking at the distances between the 12 and 13-hour points and comparing it to the distance between the 19 and 20-hour points. Schöner began the construction of the diagram by laying out the large quarter circle and drawing the various angles for the latitude values from the origin located at the intersection of the diagonal line and the x–y axis of the quarter circle. The angles then provide him with a precise measure of the variable spacing between the degree markings that can be seen on the left hand (y axis) of the diagram. While it is impossible to determine how he actually constructed the figure, he found a rich group of sources for the numerical information that it contains in the many Ptolemaic atlases in his possession.

We can only speculate on what Schöner may have used the calculator for and why he bound it in the back of the 1482 Ptolemy, but it would have been an extremely useful tool for any globemaker or cartographer in the early sixteenth century. The fact that Ptolemy still held a prominent place in the geographical imagination of the period meant that in order to keep current with the latest information one would have had to go back and forth between new information provided by cartographers like Waldseemüller and the traditional forms of latitude found in the *Geographia*. Many modern cartographers of Schöner's time, like Waldseemüller, included both the equally spaced lines of latitude, and the more astronomically founded Ptolemaic ones, on their maps as reference.

Schöner's diagram represents just one of the innovations found in his marginal annotations and is quite clearly an attempted solution to one of the many conceptual problems that beset early sixteenth-century cartographers as they wrestled to improve upon the theoretical cartography bequeathed to them from Ptolemy.

Besides annotating his copies of Ptolemy, Schöner also annotated some of the maps he owned and used. In the *Sammelband* one can clearly see that he drew red grids over the surface of both the 1507 and 1516 maps. The grid on the 1516 *Carta marina* covers the entire surface of the map and corresponds in its alignment to the latitude and longitude scale that Schöner also drew in its margins. This grid performs an early type of georectification, and is necessary in the case of the 1516 map because it is in the form of

a sailing chart and does not contain a latitude and longitude scale.

Schöner's grid gives us some hints on how he might have used his maps as they seem to imply that he was interested in them for more than just graphic representations of the known world. For Schöner, maps showed the actual locations of places and were not simply cosmographies. The conceptual difference is critical in understanding how he used the Waldseemüller maps that he preserved in the *Sammelband*. Even Ptolemy recognized the important difference between the various scales at which one could look at a map. Geography and cosmography, for him, were descriptions of the whole known world, and were to be contrasted with chorography, the description of a region, and topography, a description of a single place.[41]

In order to make his globes and to cast his horoscopes Schöner needed accurate locations for places and he treated his maps as tools and strove to find the most accurate way to extract the geographical information they contained. How he used these coordinates for the construction of his globes and as locations for the building of astrological charts or for astronomical calculation is a question that we turn to in the following chapters.

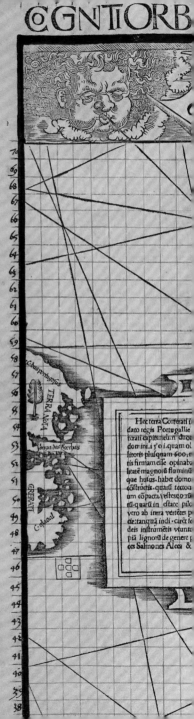

Figure 54
d lines on the *Carta marina*

höner annotated many of his maps with grid lines
order to be able to transfer coordinates from them
his globes and for his astrological predictions.

Figure 55
Schöner's scale annotations

Because Schöner used his maps as storehouses
of locations for cities and towns, rather than just
cosmographical representations, he drew scales
on them in order to determine the locations of the
various places he was interested in.

CHAPTER 2

JOHANNES SCHÖNER

Makes a Globe

Figure 69 (detail)

Figure 56
Close-up of celestial globe fragments

Figure 57
Close-up of terrestrial globe fragments

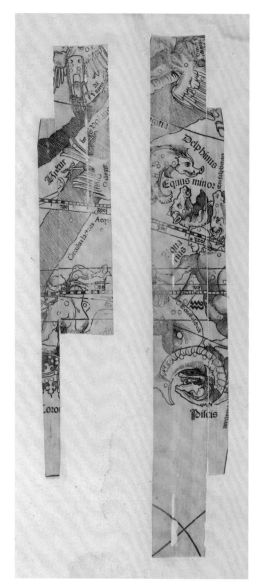

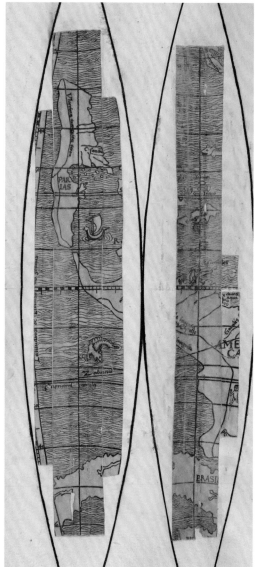

mong the most intriguing things about the collection of materials that make up the *Schöner Sammelband* are the patterns for celestial globes and the fragments of both terrestrial and celestial gores found in the binding. The fragments, which are printed on vellum, come from globes of Schöner's design and were recovered from the binding of the *Sammelband* when it was taken apart to make the facsimile of the two Waldseemüller maps for the 1903 edition by Father Joseph Fischer.

Schöner's celestial gores, although meant to form a globe, are also star charts of some complexity. Between the early fifteenth and the early seventeenth centuries, pre-telescopic star charts progressed from mere representational images, mostly based on medieval forms, to sophisticated map projections with a formalized system of stellar nomenclature.[42] Many factors accounted for this improvement, from the increasing accuracy of observational instruments, to the reintroduction of Ptolemy's *Almagest* and the constellation maps of the Islamic astronomer Umar ibn al-Sufi into the west. Instrumentally, the improvement in astrolabes also influenced the development of the star chart by providing the schematic and projective model for their display through the use of the stereographic projection. The origin of these types of charts, like most of the geography and astronomical science of the early sixteenth century, finds its foundations in Ptolemy. In Ptolemy's *Almagest* there is a catalogue of the coordinates and brightness of more than one-thousand stars, compiled from a combination of Ptolemy's own observations and historical observations by astronomers like Hipparchus. The catalogue of fixed stars has been the object of much scholarly study and Ptolemy's treatment of their motion, found in the seventh and eighth chapters of the *Almagest*, puts forward a unified theory that would dominate astronomy for more than 1,300 years.[43] Using his own observations, Ptolemy compares the alignments of the stars with those of Hipparchus and finds no measurable difference. For him, this immediately proved that the stars do not move relative to one another, but rather as a group. Ptolemy writes that,

> If one were to match the above alignments too against the diagrams forming the constellations on Hipparchus' celestial globe, he would find that the positions of the [stars] on the globe resulting from the observations made at the time of Hipparchus, according to what he recorded, are very nearly the same as at present.[44]

Ptolemy in his catalogue of stars classified the brightness or magnitude by six different numerical values, with the lesser value being the brighter star.

By the time of Schöner's celestial globes and the star charts of Albrecht Dürer, one of which he also bound into the *Sammelband*, the form of star charts was quite fixed and employed two hemispheres, one for north and one for south, on some form of polar projection.[45]

The artist and printmaker Albrecht Dürer was the first person to publish what today might

Tabula motuum stellarum quincz medio/rum longitudinis et diuersitatis.

Motus Saturni in annis coniunctis longitudinis et diuersitatis.

ħ ħ

Anni collecti	Longitudo gtes	m	2	3	4	5e	6e	Diuersitas gtes	m	2	3	4	5e	6e
18	220	1	10	57	9	4	30	135	36	14	39	11	30	0
36	80	2	21	54	18	9	0	271	12	29	18	23	0	0
54	300	3	32	51	27	13	30	46	48	43	57	34	30	0
72	160	4	43	48	36	18	0	182	24	58	36	46	0	0
90	20	5	54	45	45	22	30	318	1	13	15	57	30	0
108	240	5			54	27	0	93	37	27	54	9	0	0
126	100	8	16	40	3	3	10	229	13	42	34	20	30	0
144	320	9	27	37	12	36	0	4	49	57	13	32	0	0
162	180	10	38	34	21	40	30	140	26	11	52	43	30	0
180	40	11	49	31	30	45	0	276	2	26	31	55	0	0
198	260	13	0	28	39	49	30	51	38	41	11	6	30	0
216	120	14	11	25	48	54	0	186	14	55	50	18	0	0
234	340	15	22	22	57	58	30	322	51	10	29	29	30	0
252	200	16	33	20	7	3	0	98	27	25	8	41	0	0
270	60	17	44	17	16	7	30	234	3	39	47	52	30	0
288	280	18	55	14	25	12	0	9	39	54	27	4	0	0
306	140	20	6	11	34	16	30	145	16	9	6	15	30	0
324	0	21	17	8	43	21	0	280	52	23	45	27	0	0
342	220	22	28	5	52	25	30	56	28	38	24	38	30	0
360	80	23	39	3	1	30	0	192	4	53	3	50	0	0
378	300	24	50	0	10	34	30	327	41	7	43	1	30	0
396	160	26	0	57	19	39	0	103	17	22	22	13	0	0
414	20	27	11	54	28	43	30	238	53	37	1	24	30	0
432	240	28	22	51	37	48	0	14	29	51	40	36	0	0
450	100	29	33	48	46	52	30	150	6	6	19	47	30	0
468	320	0	44	45	55	57	0	285	42	20	58	59	0	0
486	180	3	51	45	3	1	30	61	18	35	38	10	30	0
504	40	3	6	40	14	6	0	196	54	50	17	22	0	0
522	260	4	17	37	23	10	30	332	31	4	56	33	30	0
540	120	5	28	34	32	15	0	108	7	19	35	45	0	0
558		6	36	39	41	19	30	243	43	34	14	56	30	0
576	300	7	50	28	10	24	0	19	19	48	54	8	0	0
594	60	9	1	25	49	28	30	154	56	3	33	19	30	0
612	220	40			8	33	0	290	32	18	12	31	0	0
630	140	41	23	20	17	37	30	66	8	32	51	42	30	0
648	0	42	34	17	26	42	0	201	44	47	30	54	0	0
666	220	43	45	14	35	46	30	337	21	2	10	5	30	0
684	80	44	56	11	44	51	0	112	57	16	48	16	0	0
702	40	46	7	8	53	55	30	248	33	31	28	28	30	0
720	160	47	18	6	3	0	0	24	8	46	7	40	0	0
738	20	48	29	3	12	4	30	159	46		46	51	30	0
756	240	49	40	0	21	9	0	295	22	15	26	3	0	0
774	100	50	50	57	30	13	30	70	58	30	5	14	30	0
792	320	51	1	54	39	18	0	206	34	44	44	26	0	0
810	180	52	12	51	48	22	30	342	10	59	23	37	30	0

Tabula motuum stellarum quincz medio/rum longitudinis et diuersitatis.

Motus Saturni in annis expansis et horis longitudinis et latitudinis.

ħ ħ

Anni expansi	Longitudo gtes	m	2	3	4	5e	6e	Diuersitas gtes	m	2	3	4	5e	6e
1	12	1	3	23	56	30	15	347	32	0	48	50	38	10
2	24	26	47	53	1	0	30	335	4	1	37	41	16	40
3	6	30	11	49	31	50	40	322	36	2	26	31	55	0
4	48	53	35	46	2	1	0	310	8	3	15	22	33	20
5	60	56	59	42	32	31	11	297	40	4	4	13	11	40
6	73	20	23	39	3	1	30	285	12	4	53	3	50	0
7	85	43	47	35	33	41	45	272	44	5	41	54	28	20
8	97	47	11	32	4	2	0	260	16	6	30	45	6	40
9	110	30	25	28	34	32	45	247	48	7	19	35	45	0
10	122	13	59	26	5	2	30	235	20	8	8	26	23	10
11	134	27	23	21	35	32	45	222	52	8	57	17	1	40
12	146	40	47	18	6	3	0	210	24	9	46	7	40	0
13	158	54	11	14	36	33	15	197	56	10	34	58	18	20
14	171	7	55	11	7	3	30	185	28	11	23	48	56	40
15	183	20	19	7	37	33	45	173	0	12	12	39	35	0
16	195	34	23	4	8	4	0	160	32	13	1	30	13	20
17	207	47	47	0	38	34	15	148	4	13	50	20	51	40
18	220	1	10	57	9	4	30	135	36	14	39	11	30	0

hore	Longitudo		m	2	3	4	Diuersitas		m	2	3	4	5e
1	0	0	1	23	48	41	0	2	22	49	19	14	19
2	0	0	2	47	37	24	0	4	45	38	38	28	38
3	0	0	4	11	26	6	0	7	8	17	57	42	57
4	0	0	5	35	14	48	0	9	31	17	16	57	16
5	0	0	6	59	3	30	0	11	54	6	36	11	35
6	0	0	8	22	52	13	0	14	16	55	55	25	54
7	0	0	9	46	40	55	0	16	39	45	14	40	13
8	0	0	11	10	29	36	0	19	2	34	33	54	33
9	0	0	12	34	18	19	0	21	25	23	53	8	52
10	0	0	13	58	7	1	0	23	48	13	12	23	11
11	0	0	15	21	55	43	0	26	11	2	31	37	30
12	0	0	16	45	44	25	0	28	33	51	50	51	49
13	0	0	18	9	33	7	0	30	56	41	10	6	9
14	0	1	10	19	33	21	0	33	19	30	29	20	28
15	0	1	15	20	57	10	0	35	42	19	48	34	47
16	0	1	20	22	10	59	0	38	5	9	7	49	6
17	0	1	25	23	44	47	0	40	27	58	27	3	26
18	0	1	30	25	8	36	0	42	50	47	46	17	45
19	0	1	35	26	32	25	0	45	13	37	5	32	4
20	0	1	40	27	56	14	0	47	36	26	24	46	23
21	0	1	45	29	20	2	0	49	59	15	43	0	42
22	0	1	50	30	43	51	0	52	22	5	3	15	1
23	0	1	55	32	7	40	0	54	44	54	22	29	20
24	0	2	0	33	31	28	0	57	7	43	41	43	40

Figure 58
Star catalogue from Ptolemy's *Almagest*

The star catalogue from Ptolemy's *Almagest* was the starting place for astronomical observations in the sixteenth century. Ptolemy used his own observations as well as historical ones to produce the catalogue, which contains over 1000 stars, along with their locations and magnitudes.

Figure 59
Schöner's diagram for the stereographic projection

One of the many schematic and graphic representations of map projections found in the *Opera mathematica*, this figure details the lines of a polar projection used for a star chart or planisphere.

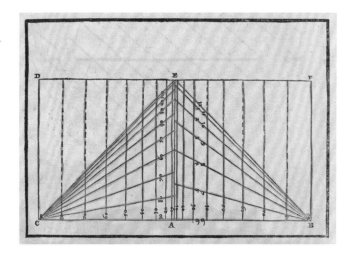

be considered an accurate star chart based on observations and plotted on paper with precision. Scholars believe that the astronomical information contained in the Dürer charts derived from two possible sources. The first is a star catalogue found in a mid-fifteenth century Viennese manuscript.[46] The second possible source is a later star chart which dates from 1503 and was produced in Nuremberg by an unknown artist.[47] The charts made by Dürer were produced in collaboration with two mathematicians, Conrad Heifogel and Johannes Stabius, and they employ a stereographic projection.

The stereographic projection, which is just one of a large number of ways of mapping a sphere onto a plane, was known by Ptolemy, who also employed it in his planisphere, and it was widely used in terrestrial mapping after 1507, when Gualterus Ludd (Vautrin Lud), one of the canons at Saint-Dié and

a friend of Waldseemüller, employed it in a small world map.[48] Schöner was very interested in the stereographic projection and produced several different forms of it along with graphic aids to its construction in his own planisphere.

The Dürer charts had an influence on all later stellar charts and globes, including Schöner's, and they were later reproduced as globes in their exact form by Gemma Frisius, in 1537.

Schöner would publish a treatise on the construction of the celestial globe that would be extremely influential for later globemakers. The book, entitled *Solidi ac sphaerici corporis, sive, Globi astronomici canones usum*, was the first to explain the techniques he developed and served as an introduction to the making of globes using printed paper gores. In his own manual on globemaking, *De principiis astronomiae & cosmographiae decque usu globi*, Frisius

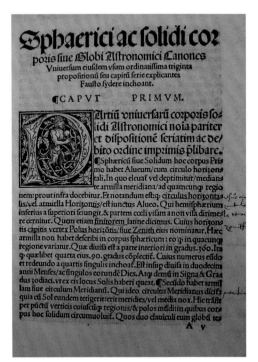

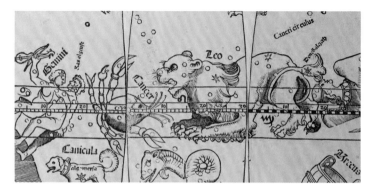

noted his debt to Schöner by granting him priority in most of the ideas expressed in it: "This is not because of the ignorance of the author or to my arrogance but it often happens that it is impossible to explain everything and it is easier to add something to things already known than to find and discover the muses."[49] Although Frisius' book contains more information on globe construction, it remains principally within the framework outlined by Schöner.

That Schöner's celestial globes were influenced by the Dürer charts is most easily seen in the symbols used to represent the stars. The stars on the gores are displayed by using two different symbols very much in line with those found on Dürer's printed chart. In both Dürer's chart and Schöner's globes, Ptolemy's six-point system of magnitudes is reduced to a system employing only two, with the vast majority of stars being represented by a simple circle, the symbol of a star reserved for only the brightest stellar bodies.

Currently four sets of Schöner celestial globes, or gores, are known to exist. Two are part of the *Sammelband* and have been shown in full earlier.

Figures 62 and 63
Tropic of Capricorn on the celestial gores and fragments

Figure 62 shows one of the missing elements on the celestial gores from the *Sammelband*. Instead of the line that should be going across the gore diagonally and that would have represented the Tropic of Capricorn, we see only the bare label. On the fragments, shown in Figure 63, on the other hand, the label not only appears but so does the line, indicating that the fragments are in a later stage of development than the full gores.

The other two sets were actually made into globes and have been dated to 1534 and 1535, based on dates found on their bases. As for the gores in the *Sammelband*, both of these sets are different, appear incomplete, and may have been proof copies. If one examines the full set of gores one notices several things missing from their design that would have been necessary for them to function as finished globes. The first missing element is that of the lines depicting the tropics. A circle, representing the Tropic of Capricorn, should be seen as a diagonal curve going across the gore, but it is missing, leaving only the woodcut label of its Latin name, *Capricorni circulus*, standing alone.

On the gore fragments the lines of the tropics are shown, as is the whole of the ecliptic, or path of the sun through the zodiac, which is the second necessary element missing from the full gores. Another unique feature of both the terrestrial and celestial gore fragments is that they are printed on vellum. Why Schöner would have done this remains a mystery, as all known gores from this period were printed on paper. The fact that Schöner discarded, and then cut up these gores, using them as guards in the binding of the *Sammelband*, may indicate that they were failed experiments.

The celestial gores from the *Sammelband* are also annotated with red numbers in selected constellations. In several of the constellations we find planetary symbols drawn by Schöner. These symbols may indicate that he was using the chart to help guide his casting of horoscopes. Schöner's numbers do not correspond to Ptolemy's numbering system, but seem

Figure 64
Schöner's annotations on
his celestial gores

Schöner annotated the gores
found in the *Sammelband* in
red ink, numbering the stars
in selected constellations and
also placing several planetary
symbols amongst them. Why
Schöner numbered the stars
in the way he did is an open
question, but the placement of
the planets may indicate that he
was recording observations on
the gores.

Figure 65
Constellation of *Gallina* on
the celestial gores

One of the unique elements of
Schöner's celestial representations is
found in the constellation of *Cygnus*.
This constellation, which should be
pictured as a swan, is instead shown by
Schöner as a rooster. Schöner names the
constellation *Gallina*, or hen, which may
be a confused translation of the Arabic
word for the constellation.

to serve simply as an accounting device for each of
the stars represented.

The artistic and mythological symbology
that Schöner uses in his representation of the
constellations shows some unique and original
forms. The constellation of *Cygnus*, the swan, for
example, is shown by Schöner as a rooster. This
confusion in the gender and species of *Cygnus*[50] is

quite common in the period before Schöner, and
many times it is simply shown as a nondescript flying
bird. The reason for the misinterpretation may reside
in the fact that in the *Almagest*, which was translated
from the Arabic, several different names are given
to this constellation. Schöner labels the constellation
Gallina, which means hen.

The sources and origins of Schöner's terrestrial
globes are more obscure and difficult to sort out
than those he made of the celestial sphere. Schöner
made his first terrestrial globe in 1515,[51] and also
wrote a short guidebook to it, called *Luculentissima
quaedam terrae totius descriptio*. The book follows in
many aspects the *Cosmographiae introductio*, written
by Mathias Ringmann and Martin Waldseemüller.
The *Cosmographiae* served as a guidebook to the
1507 World Map and the accompanying globe that
they produced at the same time. On the title page
Waldseemüller mentions the two maps that it is meant
to accompany, one *in plano* (flat) and the other *in
solido* (a globe). Waldseemüller's globe gores, of which
there are five known sets, are the first printed set of
paper gores recorded and seem to date from 1507.[52]

The fragments of Schöner's terrestrial gores
found in the *Sammelband* clearly show how much
his 1515 globe depended heavily on Waldseemüller's
map, printed in 1507, which was held into the
Sammelband using these very fragments. That map,
the first to display the New World separated from
Asia, and the first to place the name "America" on
any map, survives in only a single copy. It is one of
the great masterpieces of Renaissance cartography,
is now part of the vast map collections of the Library

Figure 66

Title page of the *Cosmographiae introductio*

The title page of the *Cosmographiae introductio* by Martin Waldseemüller and Mathias Ringmann mentioned two maps. The first is Waldseemüller's 1507 World Map and the second, a globe, of which only four copies are known to exist. Waldseemüller in 1507 was the first to print globe gores on paper, and Schöner may have been inspired by Waldseemüller's globes to make one of his own.

Figure 67

Title page of the *Luculentissima*

Much in the vein of the *Cosmographiae introductio*, Schöner produced his own guidebook to his 1515 globe. His book, while covering much the same ground, goes deeper into the geography and customs of the New World, providing detailed commentary on many locations.

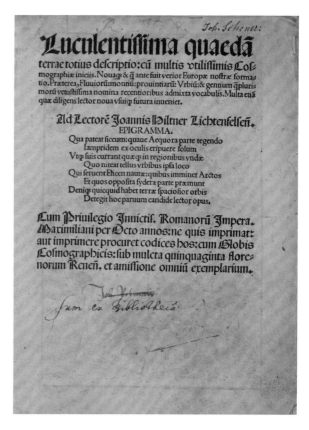

of Congress, and was of course originally part of Schöner's compilation. Waldseemüller's map was a radical departure from all previous world maps, and its creation and depiction of the world is surrounded by unanswered questions that have plagued scholars, amateurs and charlatans, almost since it was conceived in 1506.[53]

The map graphically represents a great ocean to the west of the New World, a land not yet fully explored, and which had been visited by only a few western explorers in the time since Columbus' maiden voyage of discovery in 1492. Waldseemüller's depiction of this great western ocean, years before its credited discovery by Vasco Núñez de Balboa in 1513,

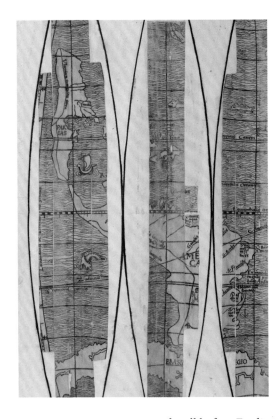

decades. All of this in an attempt to explain this map's
radical and unique vision of the world.

The outline of the continents shown on the
1507 map finds its inspiration in the great voyages
of discovery that, beginning in 1492, revealed a
new continent, full of unknown races, animals, and
myths. Waldseemüller's map is different from most
of our wayfinding maps of today: he describes it as a
cosmography, a schematic vision of what the whole
known world looked like to early sixteenth-century
geographers.

Although in form it is based on the maps of the
second-century geographer Claudius Ptolemy, the
1507 World Map is a depiction of Waldseemüller's
interpretation of the new geographical discoveries
found in both the factual and fanciful stories
emanating from the narratives and tales of the great
transatlantic explorations by navigators such as
Amerigo Vespucci. Tales of these journeys and the
otherness of the newly found lands had started to
circulate around Europe by Waldseemüller's time,
and would find perhaps their most literal expression
in Waldseemüller's other cartographic masterpiece,

and well before Ferdinand Magellan's ship reached
the Pacific Ocean in 1522, has led to speculation
of all sorts concerning how Waldseemüller might
have imagined a map that looks so much like the
modern ones we use today. Many theories have been
put forward in the four hundred years since it was
printed; scholars and amateurs alike have postulated
ideas such as secret voyages by the Spanish or the
Portuguese, and even Chinese expeditions that
trumped Columbus' discovery of the New World by

the 1516 *Carta marina*, which remains bound into the *Sammelband* as Schöner intended.

In a first glance at the fragments one is immediately struck by the similarities that they display with the 1507 World Map. The most obvious is perhaps the gap in the location of the Isthmus of Panama that is so characteristic of Waldseemüller's representation. In depicting the Pacific Ocean, one of the major sources of controversy surrounding the Waldseemüller maps, Schöner will go even further, depicting a water passage around the southern tip of South America. Unlike Waldseemüller in his later map the 1516 *Carta marina*, Schöner will retain the name 'America' on his 1515 globe.

Some of the information on the gores shows that Schöner took a keen interest in the exploration of the New World. On his gore fragments we find the path of several journeys, one in particular that has been recently identified as the voyage of Diogo Lopes de Sequeira (1466–1530), who sailed from Lisbon with four ships in 1508, and visited Madagascar, Calcutta, and Conchin.[54] Schöner's fascination with the voyages of the age of discovery and what life was like in the New World is most detailed in his guidebook to the globes, *Luculentissima quaedam terrae totius descriptio,* where he writes,

> America or Amerige is the New World and the fourth part of the earth, named for its discoverer Amerigo Vespucci, a perceptive man, who discovered it in the year 1497. In it there are brutish men who are tall and of elegant stature; they live on fish that they catch at sea.

> They have no villages of houses or huts, but only the large leaves of trees beneath which they protect themselves from the heat of the sun, but not from the rain. There are many animals of different species. They worship the heavens and the stars. In some parts of this land they have dwellings made in the shape of bells. There are red parrots and even others of a different color are found there. This island is of great size but is not wholly known. In it both the women and the men go about exactly as their mother bore them. The men there are cannibals called anthropophagi who eat their enemies.[55]

Schöner's description is quite detailed in contrast to Waldseemüller and Ringmann, who simply write,

> Today these parts of the earth [Europe, Africa and Asia] have been more extensively explored than a fourth part of the world, as will be explained later, that has been discovered by Amerigo Vespucci … The location of this part of the world and the customs of its people can be clearly understood from the voyages of Amerigo Vespucci that we have placed after this introduction.[56]

We can conclude that Schöner maintained an active interest in these discoveries throughout the period of his globemaking. One of the clearest statements of his knowledge of the history of the New World is found in a letter of dedication written in 1523, to accompany a globe he made for Reimer

von Streitberg, the Bishop of Bamberg. Because
the letter outlines Schöner's reasons and mindset
in studying new geography and his rationale for its
representation, I will quote the letter in full below:

To the most illustrious and most learned Reimer von Streitberg, Leader and Vicar in all
spiritual affairs, Father in Christ, Bishop of Bamberg and the most wonderful canon of the
Imperial Church of that City, Johannes Schöner of Carlstadt, wishes to you good health.

In as much as new inventions reconcile the minds of men to peaceful life and mitigate
unfriendly feelings, several persons of rank have urged me many times to forward to you,
your most excellent Bishop, new information that might be of interest to you and that I
have the power to do so. As I was just wondering how I might gain your goodwill, a globe
of the type that I have so often made, representing the whole surface of the spherical earth
came into my mind. It is interesting to recall that the most remote of the corners of the earth
have, in our time, been traversed by the most adventurous voyages undertaken through
the patronage of the Kings of Castile and Portugal, and it is through their patience and
generosity in defraying all costs that such explorations have been possible. Similar attempts
have of course been made in former days, but partly through the want of means, and partly
through the danger of the voyages and disasters at sea, these attempts have for the most part
been unsuccessful. No geographer has ever made in their writings about the most remote
regions and islands any mention of these new places, even though Ptolemy, by far the most
skilled of all of them, believed that he had described almost all the most distant regions of
the world. Nor did the Roman Senate itself proceed so far, for when they thought they had
subdued the whole world, they endeavored to make a survey of it.[57] In our day the King of
Portugal, after careful contemplations, directed that all necessary preparations be made and
a strong fleet equipped, and he directed they be sent out on the difficult voyage to the new
lands. Accordingly Vasco de Gama, his admiral, sailing southward and to the east, passed
over a great ocean of so large an extent that he reached the flourishing Indian provinces
of Calicut and Malaca. Meanwhile, the most worthy King of Castile, having outfitted a
fleet equally as well, dispatched his captain, Columbus to the west, and he discovered the
Antilles, Hispaniola, Cuba, and other larger inhabited countries. But so that neither of

these two great kings would have the better and take more than their fair share of the newly discovered lands, it was decreed by Alexander VI that a line should be drawn from the North Pole to the South, and that it should serve as a boundary line between the Castilian and Portuguese parts of these new lands. This arrangement was made in AD 1492.[58]

Again, in AD 1498,[59] the same King of Castile sent out Columbus to explore the great unknown regions more fully and in that voyage Columbus discovered an extensive province of huge extent, and which might very well be called a New World receiving the name Terra Firma. Columbus was unable to make further explorations of this land because of the dangerous condition of his worn ships and so he returned to Spain. The King's interest however in these new lands continued to increase and he directed that additional ships be built so if land would be found on the other side of Terra Firma, such as Paricis and Uraba, it would not escape him, for there lofty buildings and well-built towns were observed. Gold, pearls, precious stones and other valuable articles were brought to him. Still anxious to know more he dispatched a fleet of many, many ships under the command of Ferdinand Cortes in the year 1515.

For other voyages of discovery, the admiral selected was Magellan, a Portuguese, well experienced in navigation, who then set sail on the 10th August 1519, but who met with an untimely end, as did his successor John Serrano. Serrano's successor however conducted his fleet through the most remote parts of the ocean imaginable so that he sailed completely around the world in three years' time. What dangers he met are not easy to describe but will be easier for you to imagine than for me to describe at length. Having sailed all around in various directions so as to leave no portion of the route unexplored, he returned to Spain where he arrived on the 6th of December AD 1522.[60] What wonderful adventures and what extraordinary men and animals they beheld, your reverence will read at length about in the letter addressed to Maximilian, the Cardinal of Salzburg.

To add to this wonderful survey of the earth and so that what appears new and bold may appear more likely to the reader, I have at pains constructed this globe. I have copied this from a very accurate map for this that an ingenious Spaniard sent to a person of distinction. I do not however wish to set aside the globe constructed many years ago, as it fully showed all that had, at that time, been discovered. So my former globe agrees with this later one.

Please then receive this globe in the same friendly spirit in which I undertook it for your gratification. I am sure that you will not look down on my humble attempts. Farewell.[61]

Figure 69

Globe from the *De compositione de globi terrestris*

Schöner's depiction of the globe in *De compositione* is interesting in that there are some elements missing from it that would be part of later globes. Like the earlier 1492 globe of Martin Behaim, Schöner's globe does not possess an hour ring on top of the meridian line, nor does it possess an ecliptic circle.

It is obvious from the letter that Schöner had a fairly precise understanding of the Voyages of Discovery, who sponsored them, and their political ramifications.

One thing that Schöner's 1515 globe and his fragments betray is his strong interest in astrology, which will be discussed in the next chapter. On the gores, Schöner has legends indicating the kingdoms of the Three Wise Men, and also indicating the location from which the Star of Bethlehem was first seen, a theme that will come back again in his astrological writings of the 1530s.[62]

The oldest surviving terrestrial globe is that made by Martin Behaim and dates from around 1492.[63] The idea that this was the earliest globe made is, however, called into question by a manuscript that survives in Schöner's hand and appears to be a copy of an older manuscript from between 1430 and 1435.[64] The treatise, which is called the *Regionum sive civitatum distantae*, starts with a detailed set of instructions on how to make a globe. The first part of the text describes how one might inscribe the locations of places on the surface of a globe and reflects the method that Ptolemy points out in the first book of the *Geographia*, except in reverse. The method described in the manuscript is quite complex and begins by dividing up the surface of the globe into four parts by means of great circles that intersect at ninety degrees. One then divides up the equator into degrees and uses a strip of vellum tacked to the pole, and that can rotate, to mark out latitude. Once this is accomplished, placing locations on the globe is simply a matter of moving the strip to the required longitude as shown on the equator, and then moving down along the strip to the required latitude in order to mark a city's location. If Schöner's directions do date from the much earlier manuscript, as has been supposed, it is one of the few surviving records indicating a much longer tradition of globemaking than the Behaim globe might lead us to believe.

Schöner seems to have ended his terrestrial globemaking activities by 1530 but much of what he knew can be found in his *Opera mathematica*, where globemaking and geography are well represented.

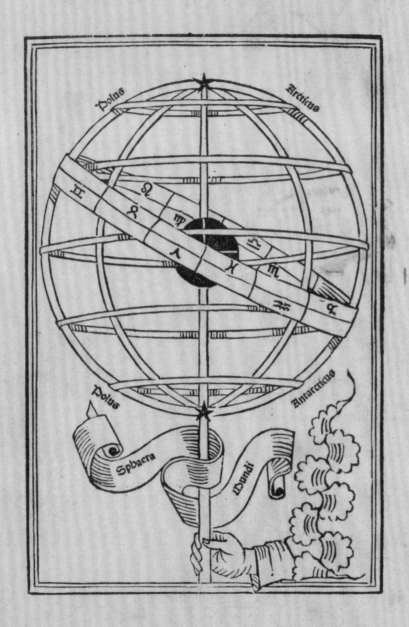

Polus Arcticus

Polus Antarcticus

Sphaera Mundi

Nouiciis adolescétib9: ad astronomicá r
impetrátib9: p breuí rectoꝛ tramite a v
Joannis de sacro busto sphęricú opusculú
planetaꝛ theoricas delyraméta Joánis d
nes tá acuratiss.qꝛ vtiliss. Necnó georgii ꝑ
planetaꝛ acuratiss. theorice: dicatú op9: v

Ractatum t
tulis disting
quid sit sphę
quid axis ꞩp
di:quot sút
múdi. ☙In
sphera mate
ꝑcęlestis qu
ni intelligit
casu signoꝛ
ctiú que sit
cis: ꞇ de diuisione climatú. ☙In q̃rto de
taꝛ: ꞇ de causis eclipsiú. Capitulum

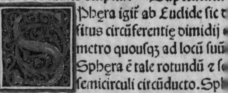

Sphera igit ab Euclide sic t
situs circúferentię dimidii
metro quousqꝛ ad locú suú
Sphera é tale rotundú ꞇ se
semicirculi circúducto. Sph
tescribit. Sphera est solidú quoddá vna
dio púct9est:a quo oés linee ducte ad cir
punct9dicit centꝛ sphe rę. Linea vo recta
plicas extremitates suas ad circúferétiá
sphere. Duo qdé púcta axé termináta t
ra aút duplicit diuidit scdm substátia:ꞇ
státiá i spheras noué scꝛ sphęrá nóná:
mobile dicit. ꞇ in sphęrá stellaꝛ fixaꝛ q

CHAPTER 3

IT'S ALL IN THE STARS

Natal Astrology and Johannes Schöner's Three Books

Figure 71 (detail)

Figure 70

Globe representation of Schöner's
Planispherum

The stereographic projection was a
subject of much mathematical study in
the early part of the sixteenth century.
Projecting the lines from the celestial
sphere onto that of a tilted earth was a
problem of some geometric complexity.
Here we see Schöner's graphic
representation of the relationship of
lines of latitude in the celestial sphere to
those on the earth.

erhaps the most unknown and least studied area of the work of Johannes Schöner concerns his books and manuscripts relating to the medieval and Renaissance practice of astrology. Schöner wrote several books on astrology during his later years and it appears all but certain that he was interested in the subject throughout his life. Astrology as a discipline was very much attached to astronomy during Schöner's lifetime and it employed the same observational methods and data, both being principally concerned with the positions of the stars and planets at particular times. The seminal texts that survived from antiquity and that became Schöner's principal sources for his astrological writings were Ptolemy's *Almagest* and his *Tetrabiblos*, both books appearing often, and in various forms, in Schöner's notes and commentary on the subject. These two texts would continue to be rich sources for astrological and astronomical knowledge even after the publication of more modern books by astronomers like Galileo, who, as is well known, taught astrology to medical students while a professor at Pisa.

The exact editions of the printed text and manuscripts that Schöner used for his notes and writings on astrology have been much more difficult to track down than some of his more obvious geographical sources, but it is worth surveying some of the possibilities before we move on to the content of the astrological works themselves. In the history of astrology and astronomy the twelfth century marks a great dividing line in the knowledge of Ptolemaic astronomical texts in the Latin west. Before the wider dissemination and translation of Ptolemy's works, most of what appeared as astronomy to scholars of late-antiquity and the very early Middle Ages was ultimately derived from compilers like Pliny or Isidore of Seville (ca. 560–636).[65] Writers like Isidore divided up the various disciplines that observed heavenly bodies into two parts, much the same way as Ptolemy did in the *Tetrabiblos*, but with a more critical eye toward astrology. In his *Etymologies* Isidore writes:

> Astronomy concerns the revolutions of the heavens, the rising, setting, motions, and etymologies of the stars. Astrology, on the other hand, is partially natural and partially superstitious. She is natural when she describes the course of the sun and moon, or certain stations of the times of the stars. The astrology pursued by the *mathematici*, however, is superstitious: they make auguries from the stars, arrange the twelve signs in the heavens according to the single parts of soul or body, and attempt to predict the nativities and morals of men from the course of the stars.[66]

In the early twelfth century, however, things changed, as a series of translations into Latin were made of Ptolemy's *Tetrabiblos*, the *Almagest*, and some of his other minor astronomical texts.[67] The first of Ptolemy's books to be translated was purely astrological, and seems to have been a form of the *Tetrabiblos*, made from Greek in 1138 by Plato of Tivoli. The use of the *Tetrabiblos* in the Renaissance, even though it went through several early printed editions, is not well understood, and while it was used as a university textbook throughout Schöner's life, the exact form of the text is still a subject of scholarly study.[68]

Shortly after the *Tetrabiblos*, in 1143, there came a translation of another of Ptolemy's works, the *Planispherum,* by Herman of Carinthia. In this text, the title of which can be translated as "celestial chart," we find Ptolemy's exploration of mapping the celestial sphere onto a two-dimensional plane, producing a kind of star chart that uses a stereographic projection much like the Dürer chart and Schöner's celestial globes that we discussed in the last chapter.

The *Almagest* itself has a much more complicated translation history than some of Ptolemy's other texts.[69] Translations of the *Almagest* were made both from Arabic and from Greek in the twelfth century and vary in their quality and emphasis.[70] The most important of these translations, and the one most widely used, was made in 1175 by Gerard of Cremona, who lived and studied in Toledo. Gerard's translation was widely circulated and greatly influenced the science of astronomy in the Latin west, especially on humanists and printers working around Schöner's time. The transmission of the translation through the Middle Ages can be traced by looking at the contents of any number of surviving early astronomical bibliographies, such as the *Speculum astronomie* compiled in the thirteenth century by Albertus Magnus. According to the *Speculum*, the most important introduction to astronomy was the *Almagest* and Albertus refers directly to Gerard's translation.[71]

The writing in the *Almagest* is extremely complicated and mathematical, and for this reason it appears to have been rarely read as a book, a situation that is very reminiscent of its reputation today. It presumes in its reader a great deal of previous knowledge of astronomy and certainly does not feel upon first reading like an introduction to the subject. Most readers of astronomy, even up until the time of Galileo in the early seventeenth century, got their astronomical education from more straightforward and less mathematical texts like the elementary *Tractatus de sphera*, the so-called *Sphere of Sacrobosco*, which contains only a few pages of detailed planetary theory.[72]

The most important astronomical textbook after the *Sphere*, and one that gave more detailed astronomical information, was the *Theoricae nouae planetarum* written by Georg Peurbach (1423–1461), based on lectures that he gave at the University of Vienna. The text became the major work consulted by most humanists interested in planetary motion. It was through Peurbach's work that they learned their astronomy and the book went through no fewer than fifty-six Latin printings during the Renaissance.[73] It was Peurbach, unsatisfied with his own book, who first began a new translation of the *Almagest,* and tried to update the inferior edition of the work by Georg of Trebizond (1395–1484). Peurbach recognized, as later did Regiomontanus, that there were serious technical problems with Trebizond's translation. Unfortunately, he had produced a draft of only the first six books when he died in 1461. After this, the project was taken up by his student Regiomontanus, who also unfortunately did not complete the work before his untimely death. And so, it was not until 1496 that a partial edition

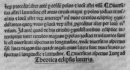

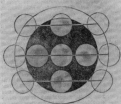

Figure 72
The orbit of Mercury

This diagram shows the orbit of Mercury as envisioned by Peurbach in the mid-fifteenth century.

of the *Almagest* was published based on the books completed by Regiomontanus.[74]

The short survey above gives some sense of the difficulties in sorting out the transmission of Ptolemy's *Almagest* to scientifically inclined humanists like Schöner at the beginning of the sixteenth century. Many of Ptolemy's other works had been printed much earlier, including texts such as the *Geographia,* which was first printed in 1475, and later in an important edition from Ulm in 1482. The *Tetrabiblos* had also by this time gone through several editions, having been printed in Venice as the *Liber quadripartitus* in 1484 and 1493. The first full printed edition of the *Almagest* would have to wait until 1515 and unfortunately it was a reprint of the old and flawed translation by Gerard of Cremona. Hence, the main text of the *Almagest*, the most important astronomical work to survive from antiquity, entered into the Renaissance mind in a translation made from Arabic based on the original Greek.[75] It was this translation and Regiomontanus' paraphrase that Schöner had available to him when he wrote his astrological works.[76]

Schöner wrote two important books that are detailed surveys of the art of astrology, and update some of the medieval forms of it that derived from Ptolemy and his commentators. Schöner's *Introduction to Astrology* (1539), and his *Three Books on Judgments of Nativities* (1545), exerted lasting influence on the field and became among the most important Renaissance guidebooks to its practice. The first title served as an introduction to the general science of astrology and astronomy, detailing as it does planetary motions, stellar locations, and the basic form of astrological symbolism.

The book in many respects has the same basic organization as is found in the early chapters of Ptolemy's *Tetrabiblos*, which Schöner relied on for much of this introductory information. In the *Tetrabiblos*, Ptolemy differentiates between two different but interconnected types of astronomy and astronomical observation. The first, which he terms astronomy proper, studies the astronomical cycles and movements of the non-fixed stars. The second, which today we would call astrology, investigates the changes these movements bring about in the terrestrial world below and tries to predict how their motions and locations effect change in the course of future events. Ptolemy writes that each of these forms has its own methodology and the former is desirable in its own right "even though it does not attain the result given by its combination with the second."

In his introduction, Schöner writes about the planetary orbs, the constellations, the seasons, the planets, and the orbits of the sun and moon in a way

IOANNIS

SCHONERI CAROLO-
STADII OPVSCVLVM ASTROLO-
gicum, ex diuersorum libris, summa cura pro
studiosorum utilitate collectum,
subnotata continens.

Instructio intelligendæ Ephemeridos.
Isagoge Astrologiæ iudiciariæ.
De electionibus communibus.
Canones succincti natiuitatum.
Tractatus integer electionum M. Lauren-
tij Bonicontrij Miniarensis.
Assertio contra calumniatores Astrolo
giæ D. Eberhardi Schleusingeri
clariss. Philosophi atcȝ Medici.

Norimbergæ apud Iohan. Petreium,
anno salutis M. D. XXXIX.

that would be understandable to a true beginner and in this way he parallels Ptolemy's treatment. In Book II he continues his basic introduction to the subject and lays out the groundwork of nomenclatures and the definitions of the astrological symbols he will employ in all his astrological writings and that are necessary to understand the more advanced *Three Books on Judgments of Nativities.*

The second book that Schöner wrote on astrology, *The Three Books on Judgments of Nativities,* is more detailed, concentrating on the particular field of natal astrology, which would remain a popular form of the science into the early seventeenth century and was even practiced and taught by Galileo. The published edition of *The Three Books* gives us, however, more than just insight into Renaissance astrology and Schöner's practice of it. It also provides us with some tantalizing hints into just how involved Schöner was in some of the more theological controversies surrounding the Protestant Reformation and how he came into contact with the Wittenberg Circle of humanists that centered around Phillip Melanchthon.

The Preface to *The Three Books* on natal astrology was written by the famous Protestant reformer Phillip Melanchthon (1497–1560) and was reprinted in every edition of the book including in Schöner's *Opera mathematica* published in 1551, four years after his death. Melanchthon, the son of a sword-maker from Bretten, would become an important humanist and follower of some of the less radical versions of Luther's reformation program. Early on in his life Melanchthon showed a talent for languages and was exposed to manuscript and linguistic studies through

his uncle, the great Hebraic and Cabalistic scholar Johannes Reuchlin (1455–1522). During his later school years he came under the tutelage of Johannes Stoeffler (1452–1531) and it was through him that Melanchthon became interested in the sciences of astrology and astronomy.

A leader in the humanistic movements that came about at the University of Tübingen, he left there in 1518, at age twenty-one, to become professor of Greek at the newly formed University of Wittenberg. According to a contemporary biographer, Joachim Camerarius (1500–1574), Melanchthon was in great demand as a scholar and turned down positions in several other universities before he decided to take up the professorship at Wittenberg.[77]

It was just at the time of Melanchthon's arrival in Wittenberg that the German Reformation was beginning to assert itself in that city in its more conservative and violent strains. The particular brand of theological uprising that was fermenting in Wittenberg, at least amongst the more militant supporters of Martin Luther, demanded not only

Figure 74

Schematic of a natal astrological diagram

Natal astrology was the form of the art that interested Schöner and that most of his writings and notes on the subject center around. These diagrams show the twelve houses of the zodiac and their effects, along with hints on how to fill in actual natal charts.

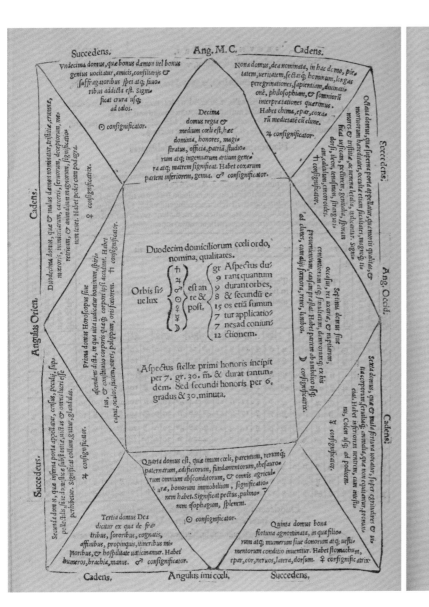

Duodecim domiciliorum cœli, sexus, complexiones, partem mundi, fortitudines, colores atq; gaudia planetarum in eis, uno intuitu capere. Canon VI.

Significator in quarta orientali A. M. C. ad horoscopum significat horas, in quarta meridiana A.M.C.usq; ad occasum significat menses, in occidentali ab occasu usq; ad I.C. significat annos, in septentrionali ab I.C.ad ortum dies.

Ptolemæo iudice partis orientalis domus sunt. { 10.11.12. / 6. 5. 4. } Occidentalis { 7. 8. 9. / 1. 2. 3. }

Me iudice partis meridionalis domus sunt. { 10.11.12. / 6. 5. 4. } Septentrionalis { 7. 8. 9. / 1. 2. 3. }

De accidentibus quæ accidunt planetis in seipsis, & ad inuicem. Progressiones siue directiones, retrogradationes, stationesq; planetarum inuestigare. Can. VII.

Vinq; illa, quæ planetis frequenter accidunt, multifariam cognosci possunt. Progreditur autem planeta, dum motus eius in ephemeride scriptus augetur. Contra uero dum minuitur, regreditur. Verum quoties nulla graduum aut minutorum motus uariatio fit, stare dicitur. Idem per syllabas Di & Re, quinq; planetarum columnulis insertas discerni potest. Præterea prima ephemeridum pagina præfata planetarum accidentia facillimè docet atq; commonstrat.

F iij Velos

a return to the simplicity of the Christian mass found in the Church's earliest and formative years, but also required the abolition of all education. The theological logic behind this movement revolved around the notion that Christ and his apostles had not been educated, and that the gospel therefore was intended for those without education. The movement theorized that the study of the Gospels needed to be returned to its more simplistic and intuitive interpretations.[78]

It was against this cultural background that Melanchthon began his career as a Professor of Greek and because of it that he decided to launch a series of educational reforms that were to change the structure and content of learning advanced in German universities. Melanchthon's reforms tried to move university studies toward a more inclusive curriculum, attempting to show how the study of nature implied the worship of Christ. Students in the sciences were instructed in arithmetic, physics, the second book of Pliny's *Natural History*, Aristotle's *Ethics*, and the *Sphere of Sacrobosco*. Those students who desired advanced degrees would have to attend lectures on Euclid and planetary theory, mostly excerpted from Peurbach and of course Ptolemy's *Almagest*. To this end Melanchthon also wrote several textbooks spanning the whole of medieval and Renaissance learning from the classical *trivium*, to physics, astronomy, history, ethics, and perhaps most influentially, Christian theology.[79]

Besides what is contained in his textbooks and curriculum reforms, Melanchthon would put forward his philosophy of education, mathematics, and the

sciences, in a series of prefaces he wrote for some of the most important scientific texts of his time. Composing prefaces to the *Sphere of Sacrobosco*, Erasmus Rheinhold's commentary on Peurbach's *New Theories of the Planets* (1542), Euclid's *Elements*, and of course Schöner's *Three Books on Judgment of Nativities*, he tried to connect the study of nature with the simplest forms of Christian doctrine. Most importantly, in his textbooks and in his prefaces he found a central role for astronomy and mathematics, specifically linking the study of these two subjects with the praise of Christ and the Creator. In his *Oration in Praise of Astronomy* (1533), Melanchthon connects the two quite explicitly:

> To recognize God the Creator from the order of the heavenly motions and of His entire work, that is true and useful divination, for which reason God wanted us to behold his works. Let us therefore cherish the subject which demonstrates the order of the motions and the description of the year, and let us not be deterred by harmful opinions, since there are some who—rightly or wrongly—always hate the pursuit of knowledge … [in the sky], God has represented the likeness of certain things in the church. Just as the moon receives its light from the sun, so light and fire are transfused to the Church by the Son of God.[80]

Writing in the preface to Schöner's *Three Books*, Melanchthon begins his justification for astrological prediction with the story of Pericles, set at the start of the Peloponnesian War. Just as Pericles was preparing to depart for Athens there arose in the sky a complete eclipse of the sun. The eclipse, according to the tale, sent fear into the helmsman of Pericles' ship, whereupon Pericles, "who either despised the threat of the stars or, pretended to be composed, covered the eyes of the helmsman with his cloak, to stop the courage of the soldier from fading." After a time Pericles removed the cloak and asked the helmsman whether anything was foretold by his eyes being covered by the cloak for the duration of the eclipse. The helmsman, who at first did not truly understand the import of the question, replied that nothing had happened, as his eyes were not injured by the cloak and that they had been covered for only a very short time. Pericles responded to the helmsman revealing the moral of his tale, "So, nor should you think that something evil has happened to the sun, or to the nature of things, or that anything is signified … by the Sun being covered for so short a time."

Melanchthon tells this tale in order to show that astrology was critical and that its predictions, if ignored, could result in tragedy, a position which had been under some attack at the beginning of the sixteenth century. He ends his story of Pericles and his justification for astrological divination by writing:

> And he ordered the setting of sail and bringing war on Greece, something that was at first extremely destructive to the nation of Athens and therefore imperiled the rest of Greece. I think, however, that there are many who might praise this judgment of Pericles, but

Figure 75
Preface by Phillip Melanchthon

This Preface by the Protestant reformer Phillip Melanchthon provides a defense of astrology and outlines its power to predict future events. Also included is a characterization of Schöner's work in the preservation of astronomical and astrological texts that would be a theme running through much of the contemporary writing about his life and work.

all of Greece suffered exceedingly disastrous penalties because of this most rash judgment.

Melanchthon uses this tale as a rhetorical device, opening up a discussion of forms of "divination" that he spoke of in his *Oration in Praise of Astronomy*. Melanchthon even compares astrological prediction with how humans have come to the knowledge of Jesus Christ, who according to him "has also revealed himself through illustrious testimonies."

Following Ptolemy, Melanchthon divides the practice of astronomy into two parts. The first division of the subject is purely observational and "Shows the true and certain laws of the motions of bodies," while the second, which he describes using the Greek word *mantike,* or divinatory art, "shows the effects or significations" that result from these motions.

The first part of astronomy Melanchthon believes to be uncontroversial as it has "utility for everyone." Nevertheless there are dissenters, "many who are either stupid or too proud and they reject this art, and by their own example lead youth away from this the most important part of philosophy." Melanchthon, in order to bolster his position, brings scriptural evidence into the discussion against these proud and stupid people, writing that,

> The judgments of these ignorant men are refuted by divine testimony. For it has been written, 'and there are lights which give us signs for the seasons, for the days, and for the years.'[81] … God has desired that the underlying principles of these signs and motions be

processed by man … Finally, let us understand that the dignity of the study of the motions encompasses many other arts (and is the origin of Geography) which we surely cannot be without in life.

The second part of astronomy, that of divination, Melanchthon, the humanist, recognizes as a point of controversy, "which many men have written against in great and many volumes." He begins his defense of astrology and of Schöner's book with a short dismissal of what was perhaps at the time of *The Three Books on Judgment of Nativities* the most profound critique of astrological prediction, that of Pico della Mirandola's *Disputations against Divinatory Astrology* (1496). Melanchthon writes that there is, "in our hands a collection of accusations from Pico not written but taken by him from ancient commentaries which were collected before the practice of divination was called into doubt."

Pico's book attacked the main predictive tenets of astrology and attempted to show how it conflicted with Aristotelian physics, common experience, and Christian dogma. In his important treatise *Oration on the Dignity of Man*, Pico would claim that a man chooses his own destiny by mastering the physical world around him, not the other way around.[82] By placing confidence in their ability to predict future events, Pico felt that astrologers were denying the free will that had been placed into individuals by the creator. He would assert in the *Disputations* that he was against astrology not because of its subject matter, but rather he doubted that its claims could be supported and "If the whole of astrology founded itself upon reasons, experiences, and authorities, and did not seek out more faithless foundations, I would not deem her wholly ruinous, but in need of support."

Schöner and Melanchthon saw it differently, of course, and found support for astrology in nature, within Christian doctrine, and in the writings of pagan philosophers. To counter Pico's collected and ancient counter-examples against astrology, Melanchthon quoted from Galen, whose reputation as a scientist and as a medical writer was at a peak in the mid-sixteenth century: "It has been shown by experience that there are effects of the stars, the sun and the moon on the elements, and on the mixed bodies, and also on plants and animals."

According to the astrologers, the effects of planetary and stellar motion were even more profound on the lives of humans, and Melanchthon reads into these effects a divine presence. Paraphrasing Aristotle in Schöner's Preface, he says that, "in the creation of human bodies, whether they are started from matter or from seed, it is known that these are strongly influenced by the power of the stars." For astronomers like Melanchthon and for astrological authors like Schöner, these effects are very much part of the causality structure of the universe, and are not seen as supernatural. For Schöner there is no need to postulate a causal connection beyond the motions and positions of astronomical objects and how they influence the happenings below; they simply exist as correlations.

Later several astrologers would begin to look for more physical effects as the foundations for these correlations and try to relate them to how the planets moved. Filippo Fantoni (1530–1591), for example, in his defense of astrology sees physical forces at work in the celestial motions that directly impact human beings:

> But let us see how they act. We should first note that celestial bodies act by motion, as is clear from the second book of *De Caelo*, where Aristotle says that heat and light are generated by air agitated from the motions of the heavenly bodies, and therefore by compressing the air they act on things below, because thus they perfect their motions.[83]

Melanchthon draws out some of these same types of correlations between astronomical positions and their effects by providing what he thinks are certain and true examples of them. He explains, for example, that "those who are born a little before conjunction, when the Moon is covered by the rays of the sun, are far more feeble than others. Again, those on whose birth the great conjunctions of Jupiter and Saturn shine forth grow much stronger."

Most of the writings of Johannes Schöner on astrology fit neatly into the mold that Melanchthon outlines in his Preface and in his two books he also attempts to justify astrological prediction in terms of a divine and Christian component. Schöner's works on astrology are also, like most of his other writings and notes in fields such as geography and mathematics, collections that preserve the work of earlier astrologers, especially those of Ptolemy and his later medieval Arabic commentators. This role of a preserver of texts

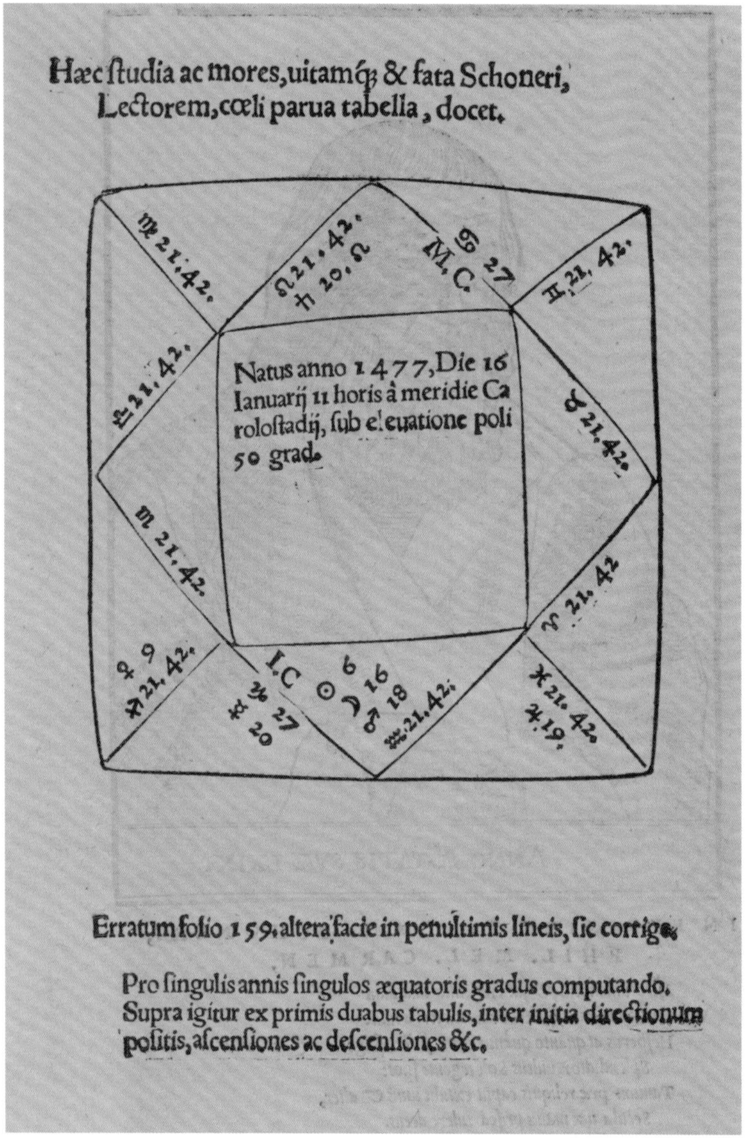

The chart contains, among other Latin text:

Hæc ſtudia ac mores, uitamq̃ & fata Schoneri,
Lectorem, cœli parua tabella, docet.

Natus anno 1477, Die 16 Ianuarij 11 horis à meridie Caroloſtadij, ſub eleuatione poli 50 grad.

Erratum folio 159, altera facie in penultimis lineis, ſic corrige.

Pro ſingulis annis ſingulos æquatoris gradus computando. Supra igitur ex primis duabus tabulis, inter initia directionum poſitis, aſcenſiones ac deſcenſiones &c.

Figure 76

Horoscope of Johannes Schöner

The Preface to Schöner's book on natal astrology, as printed in the *Opera mathematica*, ends with a natal chart of his birth.

and early scientific information is of course a theme that runs through this book and it is one that Phillip Melanchthon sees in Schöner's work as well:

Because the art of divination is natural and shows how these motions affect temperaments and inclinations, and also how the observation of celestial motions and causes can be used in medical predictions, it is an art that good minds can make use of. On this account it is important that the opinions of ancient writers be preserved. Here Johannes Schöner, a man most illustrious to me, and whom I can say that in our age to be almost the only craftsman of this kind, has put the scattered parts of this art most beautifully together so that the entire art can be learned by the studious without great effort. And Schöner, who has accomplished this through his own efforts, should not be deprived of praise on this account for otherwise the whole of this great part of natural science might perish.

Fittingly, Melanchthon's Preface to the *Three Books* ends in the *Opera mathematica* with a natal chart displaying the horoscope of Johannes Schöner.

Many writers and scientists interested in astrology would also see Schöner in the same light as Melanchthon. The Nuremberg printer Johannes Petreius, for example, writing in a letter to Schöner's student Rheticus, discusses the subject of nativities and gives more testimony to Schöner being a preserver of older texts:

> This branch of learning, which investigates the motions of the heavenly bodies, has great usefulness in all of life. And thus not only do I have an excellent opinion of you, but I also entertain great hope that it will happen that through your work much splendor will be contributed to this entire branch of learning. And I do not doubt that our fellow citizen Johannes Regiomontanus, whose many and distinguished writings our Schöner has published, employed a similar industry in his youth. When in recent days we saw this book on nativities in Schöner's library, although a little while ago we published an Arabic writer with a similar subject … we also thought this worthy to be published not only because to a virtuous man there can be no more worthy concern than that the writings of learned men be preserved, but also because this branch of philosophy concerning nativities has sure and great advantages for conducting the course of life properly without superstition.[84]

The content of the *Three Books on the Judgment of Nativities* centers on the construction of a single astrological chart, that of the Habsburg Emperor Maximilian I, which is used to illustrate the complex methods and techniques of natal astrology. Natal astrology is based on the overriding assumption that an individual's life prospects and his personality traits can be determined by looking at the positions of the sun, moon, and planets, along with the angular relationships between them at the time of a person's birth. To do it successfully, at least according to its dictates, requires that the exact time and location of a person's birth be known, along with the precise orientation of the astronomical bodies. Compiling these measurements, which are composed of latitudes and longitudes on the earth and in the heavens, are observational and cartographic tasks, which comprise most of Schöner's interests throughout his lifetime. His gathering these measurements together presents us with a possible reason why he collected the maps, atlases, and star charts that make up much of his surviving library and why he saved compilations like the *Schöner Sammelband*.[85]

The logical structure of the *Three Books* very much parallels Books III and IV of Ptolemy's *Tetrabiblos,* with Schöner's chapter headings nearly identical with those of the earlier Greek work. Schöner begins, as does Ptolemy, with a chapter on parents, then brothers and sisters, the form and figure of the body, and the qualities of the soul. One of the more important innovations found in Schöner's work, which distinguishes it from other astrological manuals of the period, is that he goes beyond Ptolemy

didiceras, eritꝗ per Animodar examinata. Nam ifto modo uerificauerunt Ptole=
mæus, Abrahamus Auenezre, Omar & alij, & non fecundũ quantitatem,continuam
fiue Geometricam,ut Alchabitius uoluit. Accipe huius hoc exemplum: Quidam
magnificus dominus natus fuit anno Domini 1459 currente die 22 Martij à meri=
die horis 4. minutis 49, non æquatis diebus. Sed æquatis diebus, à meridie horis 5,
cuius figura æftimatiua hic ponitur.

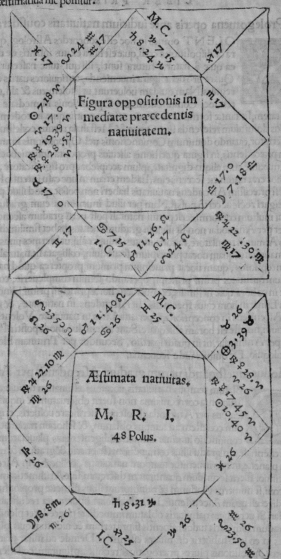

Figure 77

Natal Chart of Maximilian I

This double natal chart of Maximilian I is used by Schöner in order to instruct the reader of his *Three Books on the Nativities* in the casting of natal horoscopes. In the book Schöner outlines two methodologies, both of which are shown on this chart.

in most of the subjects he treats, including important material that derives directly from Arabic treatises on the subject.

Beginning his *Three Books*, Schöner details what it is that natal astrology consists of and what can be expected of it. He writes that:

> All astrologers are for the most part in agreement in this matter; that from the configurations and the situation in the heavens at the exact time of an infant's birth and coming from the mother's womb there can be seen things which will come to be according to nature for the person born at that time. It is for this reason that the ancients created various rules for finding the true arrangement of heaven.

These rules, all of which are highly technical in their specifics, take several hundred pages to explain, and Schöner concentrates on elucidating just two of the many possibilities. Both methods, according to Schöner, are found in the *Tetrabiblos*:

> For some, such as Ptolemy and others, the *Almuten* of the place of conjunction or opposition preceding the nativity is of such strength and potency that everything which happened in that particular month must be referred to its rulership. For this reason when Messahalla, a most ancient writer on celestial matters, found the lord of the conjunction or opposition which immediately preceded a nativity in an angle … he took it as the significator and no other planet. This was one of his great secrets …
>
> Moreover, others, such as Hermes, have proposed that the hour of the birth, called by them a second beginning, is connected naturally with the hour of conception, the first beginning, for this reason; because when nature completes the fetus working according to qualities of the first beginning, it forces the infant to go through birth in its conformity and according to its dictates. Hence he said that the place of the moon in the hour of the sperm should be ascendant in the nativity …

The technicalities expressed in the above passages are typical of astrological writing in the beginning of the sixteenth century and need not concern us as they presuppose in both the modern and Renaissance reader a great deal of both astrological and astronomical knowledge. Borrowed words like *Almuten*, from the Arabic *al-muteen*, meaning "strong in power," are found throughout Schöner's texts and betray their eastern origins.

To explain the two sets of rules or methods, Schöner sets out to create astrological diagrams or natal charts for Maximilian I using both methods. He calls the methods *Animodar* and *Trutine of Hermes*. The method of *Animodar* aims to correct the degree of the ascendant once the astrologer knows which sign that is. Ascendancy is defined as the degree of the ecliptic, or the path of the zodiac that meets the eastern horizon and is defined this way because planets located at this point will ascend above the horizon and become visible to the naked eye. This is important to know at the time of the subject's birth. *Animodar* is the form of astrology that Galileo practiced when he made up his own astrological charts in the early seventeenth century.

Schöner's astrological texts are extremely complex and I shall quote some of the first example given by him so the reader might get a feel for how difficult they are to understand:

> Take this example of the method [of investigating the true degree of the ascendant by means of the *Animodar*]: a particular high official was born in the year of the lord 1459 with the current day actually being the 22nd of March, 4 hours and 49 minutes after noon without equating the days, but equating the days 5 hours after the meridian. (This can be seen in the first natal chart.)
>
> The opposition of the luminaries

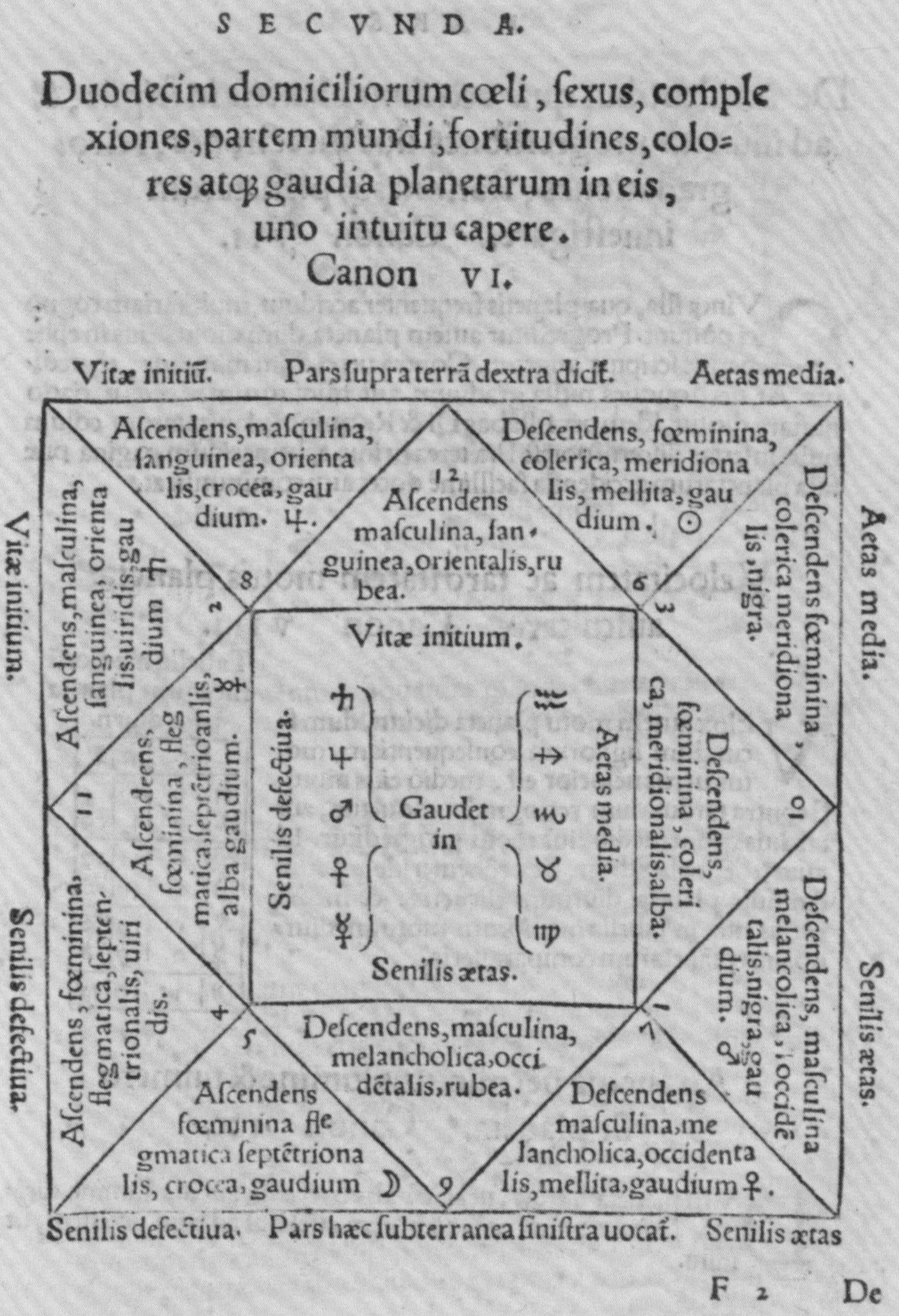

SECVNDA.

Duodecim domiciliorum cœli , fexus, comple
xiones,partem mundi,fortitudines,colo=
res atq̃ gaudia planetarum in eis,
uno intuitu capere.
Canon VI.

Vitæ initiũ.   Pars fupra terrã dextra dicť.   Aetas media.

Vitæ initium.

Afcendens,mafculina, fanguinea, orienta lis,crocea, gau dium. ♃.

Afcendens mafculina, fan= guinea,orientalis,ru bea.

Defcendens, fœminina, colerica, meridiona lis, mellita, gau dium . ☉

Afcendens,mafculina, fanguinea, orienta lis,uiridis,gau dium ♄

Defcendens fœminina, colerica meridiona lis ,nigra.

Afcendeens, fœminina , fleg matica,feptêtrioanlis, alba gaudium.

Senilis defeċtiua.

Vitæ initium.

Gaudet in

Aetas media.

Defcendens, fœminina , coleri ca, meridionalis,alba.

Senilis ætas.

Afcendens, fœminina fle gmatica,fepten= trionalis, uiri dis.

Defcendens,mafculina, melancholica,occi dẽtalis,rubea.

Defcendens, mafculina melancholica, i occidẽ talis,nigra,gau dium. ♂

Afcendens fœminina fle gmatica feptẽtriona lis, crocea, gaudium ☽

Defcendens mafculina,me lancholica,occidenta lis,mellita,gaudium ♀ .

Vitæ initium.   Senilis defeċtiua.

Senilis defeċtiua.   Pars hæc fubterranea finiftra uocať.   Senilis ætas

Aetas media.   Senilis ætas.

F 2   De

Figure 78 (left)
Diagram from the *Introduction to Astrology*

The *Introduction to Astrology* is filled with nomenclature diagrams that need to be learned by the astrologer in order to deal with what seem like infinite contingencies in casting horoscopes.

Figures 79 and 80 (opposite)
Annotations from a copy of the *Introduction to Astrology*

Schöner's astrological texts were extremely popular and very well used throughout the early sixteenth century. Most copies are heavily annotated, and some even contain original natal charts and notes on horoscopes like those shown opposite.

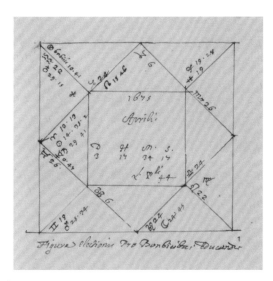

immediately preceded this nativity. The opposition was in the year 1459, which was passing away, on the 18th day of March 18 hours and 4 minutes from noon; the chart was diurnal and the place of the sun was above the earth in 7 degrees and 18 minutes of Aries. The victor, or the *Almuten*, in the place of the opposition was the sun. For this reason, seek the true place of the sun at the offered time of the nativity which is found in 10 degrees 40 minutes of Aries. However one must observe that if the moon happened to be the victrix, her true place should be found using equated days. The cause of this is the slowness of their motions, which does not allow an error worthy of correction to occur in a time of 31 minutes … However one must note that the method of the *Animodar* does not get the position in this instance such that the mid-heaven should be 10 degrees 40 minutes of Gemini. For this reason, one should use the method of *Trutine of Hermes*.

The complexities of making up a natal chart are twofold. The first is the difficulty associated with providing the astronomical observations that go into the chart, and the second is its interpretation. A natal chart, like those used by Schöner for Maximilian, is divided up into twelve parts, which are called the Houses. Each of the twelve parts corresponds to a particular constellation of the zodiac and a particular series of effects. The twelve houses represented on the chart are filled in with the locations of the planets that are in them at the time of an individual's birth.

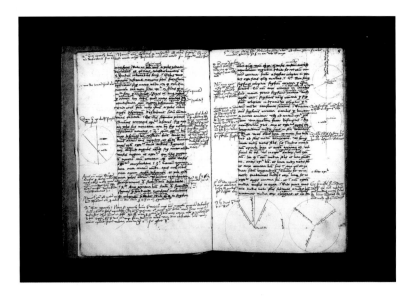

The twelve houses used in Renaissance astronomy
were very specific in providing the range of effects
that a planet might have. For example the first
house was called *vita*, or life, the second *lucrum*, or
wealth. Each of the houses represented an area of
the zodiac, like Aries for the first house, and Taurus
for the second. The planets Saturn, Jupiter, Mars,
Sun, Venus, Mercury, and the moon that moved
through the zodiac, also had specific characteristics.
For example, Saturn was considered a cold and
dry planet and if it was found in the sixth house,
valetudo, or health, at the time of a person's birth, it
could create melancholy.

These observations and calculations of the
position of planets were certainly the purpose of
astronomy and their interpretation was that of
astrology. The entire system was, however, internally
quite consistent and formal. It is due to this formality,
and to the fact that a Renaissance astrologer had
to write all of this down on paper in order to cast
a horoscope, that we now know so much about its
practice in this period.

One of the things that Schöner was most
concerned with in his writings on astrology was
the concept of prorogation, which was a specialized
technique for the prediction of life expectancy and is
based on the methodology found in the *Tetrabiblos*
(III.10). In this chapter Ptolemy will compare the
life course of a human being to an arc along the
celestial ecliptic, or the path of the zodiac. This path,
which comprises the person's expected lifespan,
starts at a particular ecliptic point (the prorogator)
where an individual's life begins and then journeys

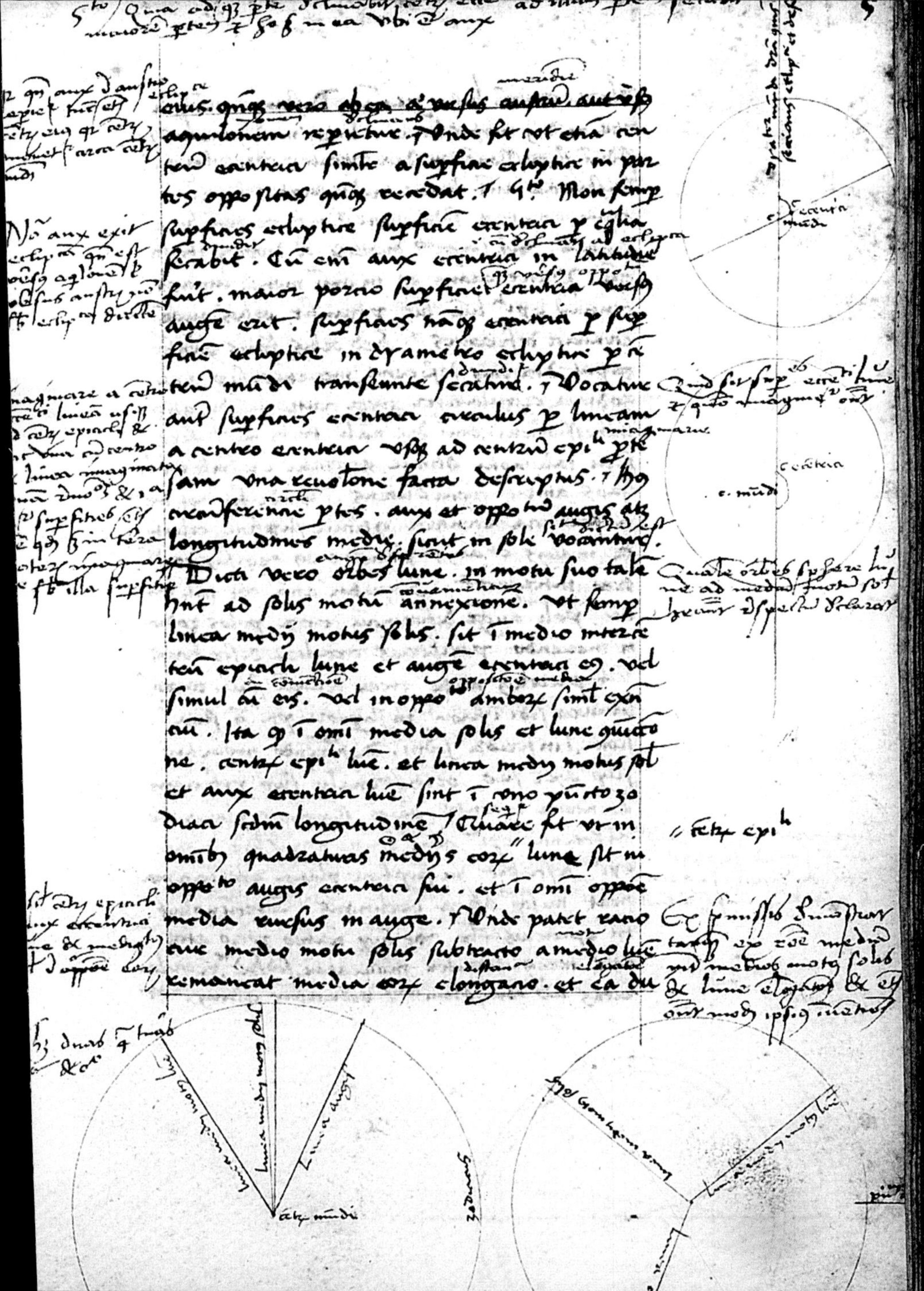

Figure 81 (detail)

along the ecliptic. Following this journey or life, and predicting its outcomes, becomes for the astrologer the astronomical problem of following the progression of that initial point, during which time it encounters any number of destructive points or planets, which represent threats, harmful events, and finally death.

In the *Tetrabiblos* Ptolemy expands the notion of the prorogator from a single point symbolizing an individual's lifespan, to no less than five possible indicators of an individual's future life course. Ptolemy explains how the horoscope, the lot of fortune, the moon, the sun, and the mid-heaven are all linked to a specific sphere of human life, which implies that every person simultaneously completes five different paths or prorogatory trajectories. He writes:

> We shall apply the prorogation from the horoscope to events relating to the body and to journeys abroad; that from the lot of fortune to matters of property; that from the moon to affections of the soul and to marriage; that from the sun to dignities and glory; that from the mid-heaven to the other details of the conduct of life, such as actions, friendships, and the begetting of children.[86]

This technique, which combined both the happening of an event and the time that it would actually happen, filled an important gap in astrological practice according to Schöner. In his *Three Books* he writes that,

> The genethliatical art consists of two things: the inquiry of things to come, on the one hand, and the examination of the time of these events on the other … As the latter teaching has little importance without the former, so the former has little importance without the latter.

This combining of event and time was an updating of the simpler methods of medieval natal astrology, which through the use of what was termed 'revolutions' were typically used to gain information about a specific future event. Prorogations of the type that Schöner concentrated on were much more powerful and allowed for the calculation of all of the critical times that might be experienced by a person through the course of their lives.

Difficult as Schöner's instructions are for us to understand today, they were certainly used a great deal during the sixteenth century to the end of the seventeenth. Most of the copies of Schöner's astrological texts that survive are heavily annotated in many different hands, with some copies containing paraphrases, study sheets, or complete natal charts bound into them.

The importance that Schöner placed in the science of astrology is obvious by the volume of writing and the amount of time he spent collecting observations and doing calculations. A greater portion of his surviving notes relate in some way to the practice of this art than anything else. Of the thirty-eight extant notebooks and collections of writings found in the Österreichische Nationalbibliothek, a full thirty-one contain astrological texts, vastly outnumbering other subjects.

In the *Three Books'* dedicatory epistle to Albrecht, Margrave of Brandenburg, Schöner clearly spelled out the importance of astrology to him, writing:

The science of the stars should be respected by us not only because of the certitude that astronomical demonstration gives, and its usefulness, but also because of its great antiquity … the sidereal science, so called by Pliny, assumes for itself the first rank because it is the one and only science which pursues the operations of the celestial parts by an admirable zeal for inquiry, and this has been done with an unstoppable diligence such that men may make pronouncements about the virtues of the stars in heaven. For the stars are not empty of virtue but are instead stronger in power and actions and greater and more beautiful than ourselves. Moreover as all the stars have different powers, so they produce actions different from each other here on earth.

He would go further and connect the practice of astrology directly to the worship of the creator:

There is no discipline that is not crippled without knowledge of the stars … The poet Virgil did not doubt that the laws of the stars kept the ignorant farmer informed about the cultivation of his fields. And at last what human condition is there on earth for which the science of the stars, praised by so many learned men through time, is not useful? No

other study makes us closer to the power and wisdom of God, the maker of all things, than contemplation of the stars above.

Astrology, as important as it was for Schöner, would not be the final resting place for his knowledge of the heavens. It was during the years when he was publishing his astrological books that he was to meet Georg Joachim Rheticus (1514–1574). Rheticus would travel from Nuremberg at Schöner's urging to Cracow and be schooled in a new type of astronomy and astronomical observation. It was from him that Schöner learned of and embraced the fact that not only did the planets move, but that the earth also traveled about the heavens. In a few short years Schöner would put his contemplation of the heavens to another use, and would become instrumental to the publication of Copernicus' *De revolutionibus*, the book that launched the scientific revolution.

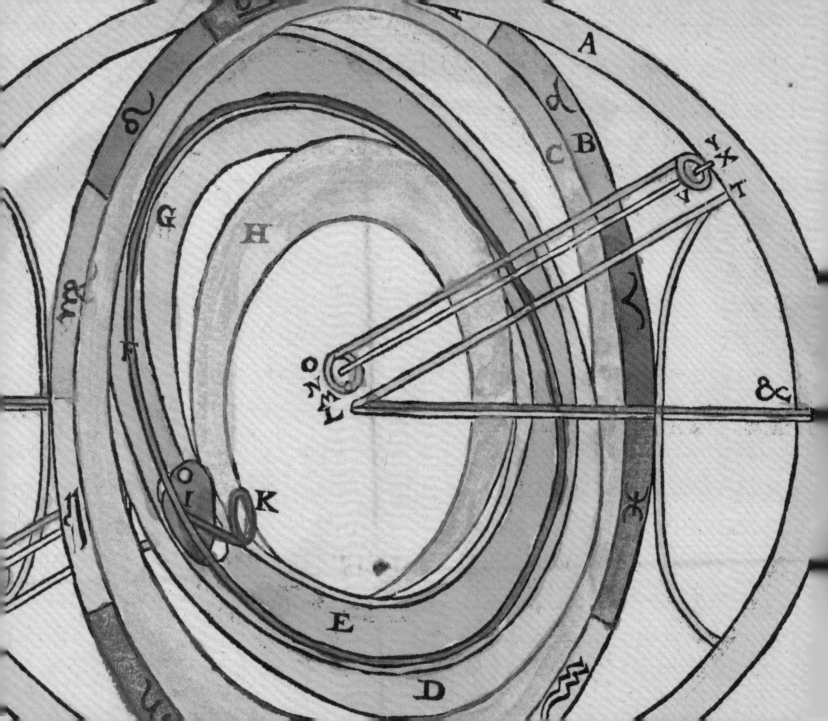

CHAPTER 4

THE EARTH BEGINS TO MOVE

Rheticus and the Early Reception of Copernican Astronomy

Figure 89 (detail)

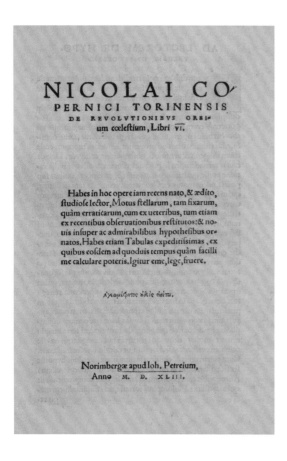

At the beginning of the first edition of *De
revolutionibus* (1543), the book that began the
modern scientific revolution and started the
earth in motion around the sun, Nicolaus
Copernicus, in a Preface dedicated to Pope
Paul II, explains to his readers why he hesitated
to publish his new theories of planetary motion.
To some extent it was simply the difficulties of the
period, with the Reformation in full swing, or, as he
says, "for fear that he might be ridiculed."[87] His fear,
it appears, centered on the unintuitive nature of his
major premise in the book that the earth revolved
on its axis and traveled about the sun. He explains in
detail that he was so apprehensive about publishing
that he had thought of giving up the project
altogether. In the dedication he further adds that it
was only because of interested friends that he decided
to finally publish the work after concealing it for a
thirty-six year period.

Although Copernicus thought about keeping
his revolutionary theory unpublished, rumors of
its radical hypothesis leaked out in the years before
1543 and circulated throughout the scientific circles
of Europe for many years. The transmission of
Copernicus' ideas came from two principal sources.
The first, now known as the *Commentariolus*,
was circulated amongst a small group of select
mathematicians and astronomers, and was not
printed in Copernicus' lifetime.[88] This short work,
which is called *Nicolai Copernici de hypothesibus
motuum coelestium a se constitutis commentariolus* in
the surviving manuscripts, contains only eight folios
and although it begins by raising objections to the

Ptolemaic system, it does not give any hints about how Copernicus arrived at the idea that the earth revolved around the sun.

In the *Commentariolus* Copernicus holds that the most important feature of a planetary theory is that the apparent motions of the planets must be made up of only uniform circular motions. This assumption is predicated on the idea that all planetary motion comes about by rotation of spheres. For Copernicus, Ptolemy's theories of planetary motion were imperfect because they allowed rotation with respect to other lines besides rotation about the center and also allowed non-uniform motion. In stressing that only uniform circular motion was possible, Copernicus was reacting not only against Ptolemy but also against the most important astronomical textbook of the period, the *Theoricae novae planetarum* of Georg Peurbach, which we have discussed in previous chapters. Peurbach's book contains very detailed and elaborate models of the spherical astronomy found in Ptolemy, but like his they violate the principle of uniform circular motion that Copernicus held as a first principle.

The *Commentariolus* contains no mathematics or geometric diagrams, but simply states seven postulates, which Copernicus goes on to discuss. The postulates are concise and go directly to the heart of his later system:

1. There is no center of all the celestial orbs or spheres.
2. The center of the earth is the center, not of the universe, but only of gravity and of the lunar sphere.

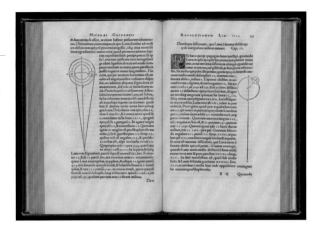

Figure 83
Geometrical diagram from *De revolutionibus*

The book was inherently mathematical and presented an updated version of Ptolemy's methods.

3. All the spheres encircle the sun, which is as it were in the middle of them all, so that the center of the universe is the sun.
4. The ratio of the earth's distance from the sun to the height of the firmament is so much smaller than the ratio of the earth's radius to its distance from the sun that the distance between the earth and the sun is imperceptible in comparison with the loftiness of the firmament.
5. Whatever motion appears in the firmament is due, not to it, but to the earth. Accordingly, the earth together with circumjacent elements performs a complete rotation on its fixed poles in a daily motion, while the firmament and the highest heaven abide unchanged.
6. What appear to us as motions of the sun are due, not to its motion, but to the motion

of the earth and our sphere, with which we revolve about the sun as any other planet. The earth has, then, more than one motion.

7. What appears in the planets as retrograde and direct motion is due, not to their motion, but to the earth's. The motion of the earth alone, therefore, suffices [to explain] so many apparent irregularities in the heavens.

The date of the *Commentariolus*, which is the first written notice of Copernicus' system, is highly problematic, and there is little evidence to support an accurate dating. What evidence there is comes from the printing history of the *Almagest* that was discussed in Chapter 3. In Copernicus' library there survives a copy of the first printed edition of the *Almagest* from 1515, which was based on Gerard of Cremona's translation. In his copy Copernicus wrote a value for the length of the lunar year that underlies a calculation of the elongation of Venus. In the *Commentariolus* he gives a very rough estimate, something that he could have found in the *Almagest* if he had had a copy nearby. This suggests the *Commentariolus* was written before Copernicus obtained a copy of the 1515 edition.[89] Besides this evidence there is a note found in an inventory of 1514 written by Matthew of Miechow (1457–1523) that states, "A manuscript of six leaves expounding the theory of an author who asserts that the earth moves while the sun stands still."[90]

The second indication and summary of Copernicus' heliocentric theory came just a few years before its publication in the form of a letter

Figure 84
The *Narratio prima*

One of the first important announcements of Copernicus' new theory came in the form of a letter written by Georg Joachim Rheticus to Johannes Schöner in 1540. The letter was first printed in Danzig that year and later appeared in the second edition of *De revolutionibus*, in 1566.

written to Johannes Schöner by his former student Georg Joachim Rheticus, in 1540. The letter, which was published in Danzig, has become known as the *Narratio prima.*

Rheticus' connection with Schöner begins when he came to study mathematics with him after he finished his degree at the University of Wittenberg. Rheticus says that he remembers, "Being attracted by the fame of Johannes Schöner in Nuremberg, who had not only accomplished much in scientific subjects, but had excelled in all the best things in life as well."[91] That the two were close is reflected in his dedication of the *Narratio prima* to Schöner, where Rheticus writes, "To the illustrious Johannes Schöner, as to his own revered father, G. Joachim Rheticus sends his greetings."[92] In the same letter, Rheticus concludes by explaining why he wanted to study with Copernicus,

> I heard of the fame of Master Nicolaus Copernicus in the northern lands, and although the University of Wittenberg had made me a Public Professor in those arts, nonetheless, I did not think that I should be content until I had learned something more through the instruction of that man. And I also say that I regret neither the financial expenses nor the long journey nor the remaining hardships. Yet, it seems to me that there came a great reward for these troubles, namely, that I, a rather daring young man, compelled this venerable man to share his ideas sooner on this discipline with the whole world.[93]

Georg Joachim Rheticus was born in 1514 and after studying in Zurich came to the University of Wittenberg. Whether through the influence of Phillip Melanchthon or through the "fame" he speaks of above, he decided to go to Nuremberg to study with Schöner. After a short time studying mathematics and geography with Schöner, he returned to Wittenberg, and began teaching mathematics in 1536.

Rheticus was however restless and, having heard reports about Copernicus' theories, he decided to travel to Frauenberg in order to see for himself. How Rheticus became informed about the new heliocentric astronomy while at Nuremberg is an open question, but it is possible that Schöner had at some time in the past come into contact with the *Commentariolus,* either directly or through the group of astronomers surrounding Phillip Melanchthon. However he became aware of the new theories, Rheticus went to see Copernicus in 1539. The collaboration between them and Copernicus' willingness to instruct Rheticus in his new astronomy is a fascinating episode in the history of Renaissance science, in that Copernicus was a Catholic canon living in a stronghold of conservative belief, hosting and teaching a man from Wittenberg, one of the most important Reformation cities in the north.

In the Preface to Johannes Werner's *De triangulis sphericus*, which was published in 1557 in Cracow, Rheticus explains that he stayed with Copernicus "three years, more or less."[94] While there Rheticus composed a short treatise, the *Narratio prima*, written in the form of a letter to his teacher Johannes Schöner. The *Narratio prima*, or 'First account', gives an intimate description of Copernicus' system. Like the *Commentariolus* it is an introduction with little in the way of geometric or mathematical details. Rheticus begins his account by telling Schöner that the rumors of Copernicus' genius and the originality of his new theory are true:

> First of all I wish you to be convinced, most learned Schöner, that this man, whose work I am now treating, is in every field of knowledge and in mastery of astronomy not inferior to Regiomontanus. I rather compare him with Ptolemy, not because I consider Regiomontanus inferior to Ptolemy, but because my teacher shares with Ptolemy the good fortune of completing, with the aid of divine kindness, the reconstruction of astronomy which he began.

Writing in an excited style, he goes on to provide Schöner with an outline of the book that Copernicus has written:

> My teacher has written a work of six books in which, in imitation of Ptolemy, he has embraced the whole of astronomy, stating and proving individual propositions mathematically and by the geometrical method.
>
> The first book contains the general description of the universe and the foundations by which he undertakes to save the appearances and the observations of all ages. He adds as much of the doctrine of sines and plane and spherical triangles as he deemed necessary to the work.
>
> The second book contains the doctrine of the first motion and the statements about the fixed stars which he thought should make it in that place.
>
> The third book treats the motion of the sun … The fourth book treats the motion of the moon and eclipses; the fifth the motion of the reaming planets; the sixth, latitudes.[95]

After his very compact outline of the contents of Copernicus' book, Rheticus advises us that he has not yet mastered all of the books. In fact he has only a good handle on the mathematics in the first three, and he only has "a general idea of the fourth and [had only] begun to conceive the hypotheses of the rest." For the remainder of the letter Rheticus details some of the evidence for Copernicus' theories, giving examples of such things as the observations of the motions of the fixed stars, the calculation of the angles of the ecliptic, and the motions of the earth and planets.

One particular group of observations, both historical and made by Copernicus of the star *Spica*, in the constellation of Virgo, were especially important to Rheticus in his letter. Copernicus uses historical observations of *Spica*, by Timocharis, Hipparchus, Menelaus, and Ptolemy to calculate the mean revolutions of the fixed stars. *Spica* would also be a star that Schöner would be interested about in his annotations to the star chart of Stabius and Dürer that he bound into his *Sammelband* with the Waldseemüller maps. Schöner annotates the chart with some of the various names of *Spica* and makes observational corrections to many of the stars found on the chart.

The question of direct contact between Johannes Schöner and Copernicus himself is complicated to unravel and rests on the series of observations that Copernicus uses in the *De revolutionibus*. In a section of the text entitled "More Recent Observations of Mercury's Motions," Copernicus lists a series of observations made using an armillary sphere. The observations of Mercury were particularly difficult and in analyzing its motion Copernicus begins,

> The foregoing method of analyzing this planet's motion was given to us by the ancients. But they were helped by clearer skies where the Nile (it is said) does not give off such mists as the Vistula does for us. We inhabitants of a more severe region have been denied that advantage by nature. The less frequent calmness of our air, in addition to the great obliquity of the sphere, allows us to see Mercury more rarely, even when it is at its greatest elongation from the sun … This planet has accordingly inflicted many perplexities and labors on us in our investigation of its wanderings.

Because of the difficulties of tracking the motions of Mercury, Copernicus tried to collect all of the observations of the planet that he could. He writes that, "I have therefore borrowed three positions from those which were carefully observed at Nuremberg. The first was determined by Bernhard Walther,

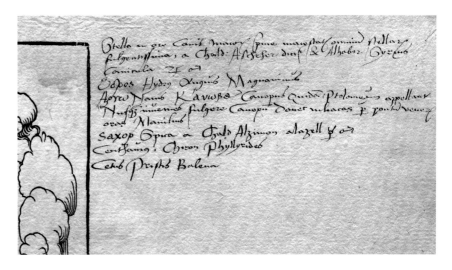

Figure 85
Annotations on Schöner's copy of the Dürer-Stabius star chart Courtesy of the Collections of Wolfegg Castle

Schöner annotated Dürer's star chart with some corrections to stellar positions and with a series of alternate names to some of the stars represented.

Schöner lists stars, such as Centaurus, and in many cases presents alternate names, like Chiron and Phyllyrides in this example. He also gives the Arabic names, Alzimon and Alazell, for the star Spica, which strangely enough is not found on the chart itself, as it resides in the northern constellation of Virgo.

Figure 86

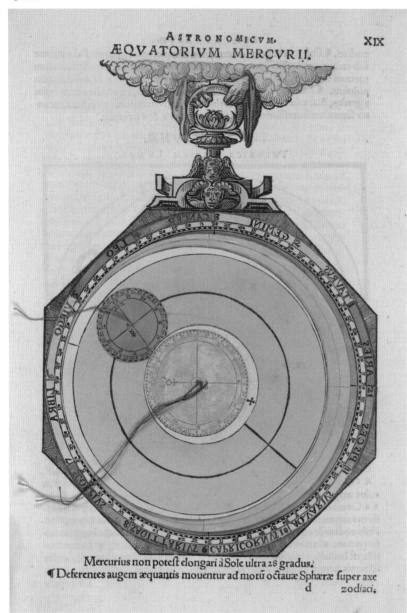

Figures 86, 87 & 89 (page 163)
The orbits of Saturn and Mercury as pictured by
Schöner in his *Aequatorium astronomicum*

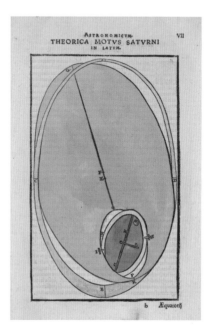

Figure 87

Regiomontanus pupil, 5 uniform hours after midnight
on 9 September = 5 days before Ides, 1491."

Bernhard Walther (1430–1504), was a wealthy
merchant and patron of the sciences, who also had
astronomical instruments built for his own use and
who made a series of observations that were used by
Copernicus. When he died his observations had not
yet been published, but his manuscripts came into the
hands of Schöner, who printed them in Nuremberg in

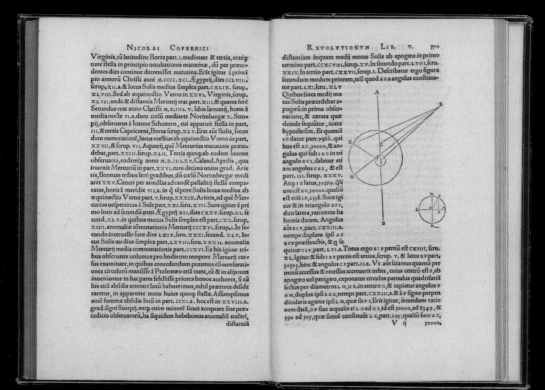

Figure 88
Mercury observations from *De revolutionibus*

Some of the most difficult observations used by Copernicus in *De revolutionibus* concern the planet Mercury. Copernicus collected observations of the planet from various sources and credits both Johannes Schöner and Bernhard Walther with several of the key positions he used in the book.

1544. The observations must have been made available to Copernicus in manuscript form either through Rheticus, or directly from Schöner himself, as they appear in *De revolutionibus* in 1543. The transmission of these observations to Copernicus is further complicated by the fact that he credits Schöner with two of them,

The second position was observed by Johannes Schöner 6.5 hours after midnight 9 January 1504, when 10 degrees of the Scorpion were culminating over Nuremberg. He saw the planet at 3.5 degrees within the Goat …

This observation was in fact performed by Walther, and mistakenly attributed by Copernicus to Schöner.[96] How Copernicus made this mistake is a matter of speculation, but it seems highly likely that it was Rheticus who brought these observations to

Copernicus' attention through Schöner, who was in all probability the person who suggested that Rheticus undertake the journey to see Copernicus in the first place. Perhaps this is the reason that Rheticus wrote to Schöner from the city of Posen telling him that he had finally decided to undertake the final leg of the trip to learn of the new astronomy. Rheticus references this earlier letter in the dedication of the *Narratio prima*, saying that "On May 14th I wrote you a letter from Posen in which I informed you that I had undertaken a journey to Prussia, and I promised to declare as soon as I could, whether the actuality answered to report and to my own expectation."[97]

What Schöner's reactions to Rheticus' letter might have been are lost to time and we have only Rheticus' words near the end of the *Narratio prima* to look to. Concluding his presentation, we can hear in Rheticus' voice the uncertainty of how the new theories of planetary motion will be received,

Most Illustrious and learned Schöner, whom I shall always revere like a father, it now remains for you to receive the work of mine, such as it is, kindly and favorably. For although I am not unaware what burden my shoulders can carry, nevertheless your unparalleled and, so to say, paternal affection for me has impelled me to enter this heaven not at all fearfully and to report everything to you to the best of my ability …

Furthermore, concerning my teacher I should like you to hold the opinion and be fully convinced that for him there is nothing better or more important than walking in the footsteps

of Ptolemy … However, when he became aware that the phenomenon, which controls the astronomer, and mathematics compelled him to make certain assumptions, even against his wishes, it was enough he thought, if he aimed his arrows by the same method to the same target as Ptolemy, even though he employed a bow and arrows of a far different type.

Rheticus ends his letter as he began it with a quotation from the *Handbook of Platonism* by the second-century philosopher Alcinous, "Free in mind must he be who desires to have understanding."

One question that remains concerning Schöner's relationship to Copernicus is the fate of his copy of *De revolutionibus*. When *De revolutionibus* was published in Nuremberg in 1543, Schöner most certainly would have been one of the first people to have obtained a copy. In the Linda Hall Library of Science in Kansas City there is a *De revolutionibus* that has led several scholars, like Ernst Zinner and the present writer, to think that it was Schöner's original copy.[98] Within the book there are a series of annotations covering a paper instrument called a *Tusi couple* that might possibly be in Schöner's hand. Owen Gingerich, however, in his masterful survey of the annotations in the surviving copies of the *De revolutionibus*, disagrees with Zinner's assessment of the same evidence, and hence the identity of the annotator remains controversial.[99]

¶Venus non poteſt elongari à Sole ultra 48 gradus. Quando autem Venus multum à Sole remouetur, & fuerit in parte inferiori epicycli, tunc ſi fuerit dies ſerena, uidebitur in die, & magna ualde noctu apparebit

¶Deferens Veneris mouetur ſingulis diebus iuxta ordinem ſignorum 59 minutis, 8 ſecundis, ſicut ipſe Sol. Argumentum autem eius in epicyclo mouetur ſingulis diebus 37 minutis fere.

THEORIA MERCVRII.

THEORICA ORBIVM MERCVRII.

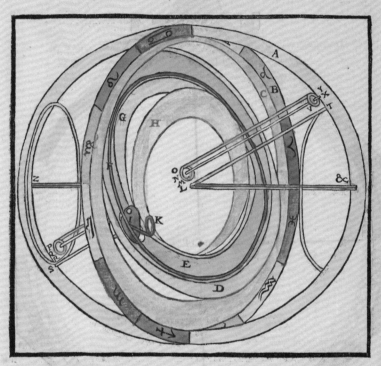

Figure 89 (detail)

¶A Circulus magnus ductus per polos zodiaci primi mobilis eiusꝗ puncta ſolſtitialia. ¶B Zodiacus primi motoris. ¶C Deferens augem æquantis. ¶D Deferens augem eccentrici. ¶E Eccentricus Mercurij. ¶F Æquans. ¶G Deferens oppoſitum augis eccentrici. ¶H Deferens oppoſitũ augis æquantis. ¶I Epicyclus Mercurij. ¶K I Axis inclinationis epicycli. ¶L Centrũ mundi.
¶M Centrum

CONCLUSION

JOHANNES SCHÖNER

And the Birth of Modern Science

Figure 90 (detail)

When *De revolutionibus* was finally published in Nuremberg in 1543, Schöner was nearing the end of his life. He would not live to see the difficulties that belief in the motion of the earth and the heliocentric system would generate for those who professed it, and who dared to move beyond the confines of Ptolemy's geocentrism. His was still a time when astronomical theories were conceived of as mere hypotheses, meant to be useful for the calculation and prediction of eclipses, the changing of the seasons, and for tracking planetary positions. Astronomical theories were useful tools, but postulated no necessary reality to the mathematical and geometric mechanisms imagined. It would not be until the advent of the telescope, more than fifty years after Schöner's death, when the ground would shift, and that an observational way of distinguishing between the Copernican and Ptolemaic systems would be possible.

As the reader is more than aware by now, it was not only in the celestial sphere that revolutionary changes were taking place. Schöner witnessed the transmission of the stories and tales of explorers recounting their adventures and the discovery of the New World. He not only saw the earth set in motion but he also saw its size expand greatly. As a globemaker, Schöner was profoundly interested in these events and collected the most modern maps available in the form of Waldseemüller's 1507 and 1516 masterpieces. We cannot be sure how Schöner perceived of this philosophically or what he thought this epistemological change meant to the larger world,

except to recognize that in his terrestrial globes he fully embraced them, creating spheres whose representation of the New World went beyond even Waldseemüller's original vision. In his geographical work Schöner seems to have taken to heart the dictum written out by Martin Waldseemüller on his 1507 World Map,

> This one request we have to make, that those who may not be acquainted with geography shall not condemn all that they see before them until they have learned what will surely be apparent to them later on, when they have come to understand what they see.

Throughout his life Johannes Schöner would always be attracted to the newest theories in mathematics, geography, and astronomy and this interest drove him to collect and save some of the most important scientific, mathematical, and cartographic materials from the ravages of time. Much of the work of Regiomontanus and Waldseemüller would never have been part of our modern cultural history had he not interceded, and if not for him we might still be like John Boyd Thacher, hoping and asking, "Where is this Waldseemüller map? … Somewhere, in some dark corner of a monastic library, folded away in some oak-bound volume a copy may be sleeping."

Unfortunately, Johannes Schöner left behind no philosophical treatise or any reflections on how he felt the theories put forward by the new advancements in science, geography, and mathematics affected his world view, and whether they revealed truths about the universe or they simply provided tools for its comprehension by men. These two poles of what science does and what the objects of its study really are remain today, as they did in Schöner's time, an ongoing series of philosophical problems. Does science uncover real facts about what the universe is actually made of, or is it simply constructed by us in an effort to understand what is around here, never actually approaching the hidden truth of nature?[100] In Schöner's time the answer to this question was easier and was perhaps put best in 1503, by the Aristotelian philosopher Lefèvre d'Étaples in his *Introductorium astronomicum*,

> This portion of astronomy is almost entirely a matter of representation and imagination. The good and wise Artisan of all things, by an act of his divine intelligence, produced the real heavens and their real movements. Similarly, our intelligence, which seeks to imitate the Intelligence to which it owes its existence, each day plotting out a little more the spots of ignorance, our intelligence I say, composes within itself some fictive heavens and fictive motions; these are images of the true heavens and true motions. And in these images, as if they were traces left by the divine Intelligence of the creator, the human intelligence seizes hold of truth. When therefore, the mind of the astronomer composes a correct representation of the heavens and their movements, he resembles the Artisan creator of all thing creating the heavens and their motions.[101]

Few scientists or philosophers would have disagreed with this in the early sixteenth century and held realist positions, thinking what they were doing was uncovering the true structure of creation; even Ptolemy did not go that far. Rather, most held one form or another of the above position, thinking that theories gave insight and allowed the calculation and prediction of astronomical phenomena, but did not make any necessary ontological commitment to the existence of the objects postulated. Ptolemy's spheres, epicycles, and equants did not really exist in the world; they were simply convenient ways to model the heavens.

But in the work of Schöner we can see glimpses of what was to come. We can see him questioning the theories of Ptolemy in astrology and astronomy, making and updating older astronomical observations, and annotating and correcting the geography of the remote past. We see in his gathering of mathematical and astronomical manuscripts recognition of the dynamism of the science of his day, and an effort to comprehend its geography through collecting the most modern maps and making globes that show how the world had changed.

No longer satisfied with what was handed down in the ancient texts of Ptolemy, and faced with the discovery of new worlds, men like Schöner began to think in a different way than their medieval and scholastic predecessors, a way that questioned and embraced the shifting intellectual climate around them. They must have felt these shifts deeply, for few things in their world, whether they were religious, scientific, or political, were standing still. It is through their eyes and through their questions that today we can see how our modern scientific world view emerged from the long Middle Ages.

The period of science between Columbus and Copernicus is a difficult one to understand and suffers greatly from a lack of scholarly attention, especially in the English-speaking world. Locked as it is into complex and technical Latin texts on astronomy, mathematics, and astrology, its accessibility to those interested in its history is limited, and reserved for the few with the combination of linguistic and mathematical tools necessary to read its primary sources and crack its codes. It is my hope that this book has changed some of that and provided a short introduction to the inner world of one sixteenth-century scientifically inclined humanist, who was trying to understand the radical and revolutionary times he was living through, and how by preserving and saving the manuscripts and maps that he did, he has helped us to understand a bit more of ours.

Figure 90
Schöner's *Planispherum*

This beautiful image, taken from the *Opera mathematica* of Johannes Schöner, presents a moveable *planispherum* that allows the reader to chart the path of the stars across the sky over the western hemisphere.

Explicandi

ENDNOTES

1. John Boyd Thacher, *The Continent of America: Its Discovery and its Baptism* (New York: W. E. Benjamin, 1896), 151, 157.

2. Several recent books tell the story of Waldseemüller's maps. For the most up-to-date research and facsimiles of Waldseemüller's maps, see John Hessler and Chet Van Duzer, *Seeing the World Anew: The Radical Vision of the 1507 and 1516 World Maps of Martin Waldseemüller* (Washington, DC: Library of Congress and Levenger Press, 2012). John Hessler, *The Naming of America: Martin Waldseemüller's 1507 World Map and the Cosmographiae introductio* (Washington, DC, and London: Library of Congress and D Giles Limited, 2008) gives a translation of the *Cosmographiae introductio* with commentary. Peter Dickson, *The Magellan Myth: Reflections on Columbus, Vespucci and Waldseemüller* (Mount Vernon, Ohio: Printing Arts Press, 2007) puts forth the theory that Waldseemüller and other early sixteenth-century cartographers knew of the Pacific Ocean and the Straits of Magellan before their accepted discovery dates of 1513 and 1522. Toby Lester, *The Fourth Part of the World: An Astonishing Epic of Global Discovery, Imperial Ambition and the Birth of America* (New York: Free Press, 2010) presents the full story and context of the Waldseemüller maps as reflected in the geographical discoveries of the late fifteenth and early sixteenth centuries.

3. Karl Heinz Burmeister, *Georg Joachim Rhetikus, 1514–1574, Ein Bio-bibliography* (Wiesbaden: Guido Pressler Verlag, 1967), vol. III, 50.

4. James Steven Byrne, "A Humanist History of Mathematics? Regiomontanus' Padua Oration in Context," *Journal of the History of Ideas* 67 (2006): 41–61.

5. B. R. Goldstein. "The Arabic Version of Ptolemy's *Planetary Hypothesis*," *Transactions of the American Philosophical Society* 57 (1964): 3–55.

6. For more on the story of Fischer's discovery, see Toby Lester, *The Fourth Part of the World* (New York: Free Press, 2009), 12–19.

7. ÖNB, Codex, 4988.

8. Johannes Schöner, *Opera Mathematica Ioannis Schöneri in vnvm volvmen congesta…* (Norinbergae: I. Montani & Neuberi, 1551).

9. There have been few studies of Schöner's library and manuscripts. An exception is a paper by Monika Maruska, "Die Handschriften aus der Bibliothek des fränkischen Gelehrten Johannes Schöner in der Österreichischen Nationalbibliothek," in *Aspekte der Bildungs-und Universitätsgeschichte*, Wien (1993): 408–31.

10. On folio 124 of Codex 3292 ÖNB, Vienna, Schöner writes that he purchased the book on the 16th October 1507. For more on Schöner's 1515 Globe, see Chet Van Duzer, *Johann Schöner's Globe of 1515: Transcription and Study* (Philadelphia: American Philosophical Society, 2010).

11. MS. Vind 3505, ÖNB. For a transcription of this text, see Dana Bennett Durand, *The Vienna-Klosterneuberg Map Corpus of the Fifteenth Century: A Study in the Transitions from Medieval to Modern Science* (Leiden: E. J. Brill, 1952), 364–67.

12. For a modern take on Galileo's life and relationship to the science and politics of his day, see J. L. Heilbron, *Galileo* (Oxford: Oxford University Press, 2010).

13. Joannis Regiomontanus, *Doctissimi viri et mathematicarum disciplinarum eximii professoris Ioannis de Regio Monte De triangulis omnimodis libri quique* (Norimbergae: Petriei, 1533).

14. ÖNB, Codex, 5203.

15. Menso Folkerts, "Regiomontanus' Role in the Transmission of Mathematical Problems," in *From China to Paris: 2000 Years Transmission of Mathematical Ideas*, ed. Yvonne Dold-Samplonius (Stuttgart: Franz Steiner Verlag, 2002), 411–28; "Regiomontanus' Role in the Transmission and Transformation of Greek Mathematics," in *Tradition, Transmission, Transformation: Proceedings of Two Conferences on Pre-Modern Science Held at the University of Oklahoma*, ed. F. Jamil Ragep (Leiden: E. J. Brill, 1996), 89–113.

16. *De triangulis omnimodis*, 2.

17. Lynn Thorndike, "Of the Cylinder Called the Horologe of Travellers," *Isis* 13 (1929–30): 51–52.

18. Elly Dekker and Kristen Lippincott, in their article "The Scientific Instruments in Holbein's Ambassadors: a Re-examination," *Journal of the Warburg and Courtauld Institutes* 62 (1999): 93–125, describe many of the same instruments made and written about by Schöner. It is interesting that the terrestrial and celestial globes found in

that painting have also been found to be related to those Schöner created.

19. Johannes Schöner, *Coniectur odder ab nemliche auslegung Joannis Schöners uber den Cometen so im Augustmonat des M.D.XXXI* (Nuremberg, 1531).

20. Lynn Thorndike, "Franco of Polonia and the Turquet," *Isis* 36 (1945): 6–7.

21. Nicolas Jardine, "Epistemology of the Sciences," in *The Cambridge History of Renaissance Philosophy*, ed. Charles B. Schmitt (Cambridge: Cambridge University Press, 1988), 685–711.

22. Ibid., 685.

23. Vienna Codices, 2992 and 3292.

24. Florian Mittenhuber, "The Tradition of Texts and Maps in Ptolemy's Geography," in *Ptolemy in Perspective: Use and Criticism of his Work from Antiquity to the Nineteenth Century*, ed. Alexander Jones (Berlin: Springer-Verlag, 2010), 95–199.

25. Much of my history of the text of the *Geographia* is taken from the seminal article by Patrick Gautier Dalché, "The Reception of Ptolemy's Geography (End of the 14th to Beginning of the 16th Century)," in *History of Cartography*, Vol. 3, ed. David Woodward (Chicago: University of Chicago Press, 2009), 287.

26. Phyllis Walter Goodhart Gordan, *Two Renaissance Book Hunters: the Letters of Poggius Bracciolini and Nicolaus De Niccolis* (New York: Columbia University Press, 1974), 9, 84.

27. Translation of manuscript Bibliothèque Nationale de France, français 9558, folio 9v as found in Dalché, "The Reception of Ptolemy's Geography," 306.

28. For more on the translation see James Hankins, "Ptolemy's Geography in the Renaissance," in *The Marks in the Fields: Essays on the Use of Manuscripts*, ed. Rodney G. Dennis and Elizabeth Falsey (Cambridge, Mass.: Harvard University Press, 1992), 119–27.

29. Dalché, "The Reception of Ptolemy's Geography," 340. Quoting from manuscript Basel, Universitätsbibliothek, O IV 32.

30. Hessler and Van Duzer, *Seeing the World Anew*, 27.

31. *De triangulis omnimodis*, 4.

32. ÖNB, Codex 3292. The inscription reads: *Emptus anno 1507 die 16 octobris hora 2a post meridiem per II flor.*

33. ÖNB, Codex 3292.

34. Much of my treatment of the technical aspects of Ptolemy's projections comes from the excellent translations and commentary of J. Lennart Berggren and Alexander Jones, *Ptolemy's Geography: an Annotated Translation of the Theoretical Chapters* (Princeton: Princeton University Press, 2000), 35–40 and 86–89.

35. For a survey of Greek mathematical and geometric methods of the type employed by Ptolemy, see Ian Mueller, "Euclid's Elements and the Axiomatic Method," *British Journal for the Philosophy of Science* 20 (1969): 289–309.

36. Ptolemy, *Almagest*, translated and annotated by G. J. Toomer (Princeton: Princeton University Press, 1998), 75–76.

37. This is the elevation of the north or south celestial pole above the horizon.

38. An upright stick.

39. I thank Owen Gingerich of the Harvard Observatory for helpful discussions regarding the diagram and its possible relationship to latitudinal calculations.

40. *Ptolemy's Geography*, 84.

41. For more on this, see F. Lukermann, "The Concept of Location in Classical Geography," *Annals of the Association of American Geographers* 51 (1961): 194–210.

42. Anna Friedman Herlihy, "Renaissance Star Charts," in *The History of Cartography: Cartography in the European Renaissance*, ed. David Woodward (Chicago: University of Chicago Press, 2007), 99–134.

43. Gerd Grasshoff, *The History of Ptolemy's Star Catalogue* (Berlin: Springer-Verlag, 1990).

44. Claudius Ptolemy, *Almagest*, trans. and annotated by G. J. Toomer (Princeton: Princeton University Press, 1984), 138.

45. Ibid, 99.

46. ÖNB, Codex 5415.

47. Deborah J. Warner, *The Sky Explored: Celestial Cartography, 1500–1800* (New York: A. R. Liss, 1979), 74.

48. John P. Snyder, *Flattening the Earth* (Chicago: University of Chicago Press, 1993).

49. Gemma Frisius, *De principiis astronomiae & cosmographiae* (Antwerp: Johannis Steelsius, 1553), 29–30.

50. Dekker and Lippincott, "The Scientific Instruments in Holbein's Ambassadors," 103.

51. The seminal study on Schöner's 1515 globe and its relationship to Waldseemüller and the Voyages of Discovery is Chet Van Duzer's *Johannes Schöner's Globe of 1515: Transcription and Study* (Philadelphia: American Philosophical Society, 2010). My interpretation of his terrestrial globes follows his research closely.

52. Hessler, *The Naming of America*.

53. Several recent books tell the story of

Waldseemüller's maps: see note 2.

54. For a complete description of this journey and Schöner's representation of it see Van Duzer, *Johannes Schöner's Globe of 1515*.

55. Quoted from the transcription and translation of Van Duzer, *Johannes Schöner's Globe of 1515*.

56. Quoted from Hessler, *The Naming of America*, 101.

57. Interesting reference to the taxation survey of Augustus.

58. Schöner is referring to the Papal Bull of 1493, which began the process towards the 1494 treaty of Tordesillas between the Spanish and the Portuguese.

59. This is the traditional date of Columbus's third voyage.

60. This actually happened in September of 1522.

61. For the Latin text and a slightly different translation, see Henry Newton Stevens, *Johannes Schöner, professor of mathematics at Nuremberg, a reproduction of his globe of 1523, long lost, and his dedicatory letter to Reymer von Streyperck* (London: H. Stevens, 1888).

62. Van Duzer, *Johannes Schöner's Globe of 1515*.

63. E. G. Ravenstein, *Martin Behaim, his Life and his Globes* (London, 1908).

64. Durand, *The Vienna-Klosterneuberg Map Corpus of the Fifteenth Century*, 164.

65. For more on the Latin history of Ptolemy's texts, see Olaf Pedersen, *A Survey of the Almagest*, Sources and Studies in the History of Mathematics and Physical Sciences (Berlin: Springer-Verlag, 2010).

66. Isidore of Seville, *Etymologiarum sive originum libri XX*, ed. W. M. Lindsay (Oxford: Oxford University Press, 1911), Chapter III.27.

67. F. J. Carmody, *Arabic Astronomical and Astrological Sciences in Latin Translation: a Critical Bibliography* (Berkeley: University of California Press, 1956).

68. H. Darrell Rutkin, "The Use and Abuse of Ptolemy's Tetrabiblos in Renaissance and Early Modern Europe," in *Ptolemy in Perspective*, ed. Alexander Jones (New York: Springer-Verlag, 2010), 135–49.

69. C. H. Haskins, *Studies in the History of Medieval Science* (Cambridge, Mass.: Harvard University Press, 1924).

70. In my commentary on the translation I use the summation of Haskins's works found in Pedersen, *A Survey of the Almagest*. There were at least four translations made. The first, dated about 1160, was made in Sicily directly from the Greek, by an anonymous translator. The most complete manuscript is Vatican Lat. 2056 and is the earliest known Latin version of Ptolemy's astronomy. Other translations include a fragmentary version most likely made from the Arabic in Spain in the early thirteenth century and a version containing only the first four books, which exists in only a single manuscript. For more details on the manuscripts, see Haskins, *Studies in the History of Medieval Science*.

71. Pedersen, *A Survey of the Almagest*, 17.

72. For more on this, see Lynn Thorndike's survey, *The Sphere of Sacrobosco and its Commentators* (Chicago: University of Chicago Press, 1949) and Olaf Pedersen, "The Theorica Planetarum Literature of the Middle Ages," *Classica et Medievalia* 23 (1962): 225–32.

73. Ernst Zinner, *Leben und Wirken des Johannes Müller von Königsberg genannt Regiomontanus* (Munich: 1938).

74. *Epytoma Joannis de Monte region in Almagestum ptolemei* (Venetiis: Hartmann, 1496).

75. *Almagesta Cl. Ptolemei Pheludiensis Alexandrini, astronomorum principis* (Venetiis: in officina Petri Liechtenstein, 1515).

76. Noel M. Swerdlow, "Regiomontanus on the Critical Problems of Astronomy," in *Nature, Experiment and the Sciences: Essays on Galileo and the History of Science in Honour of Stillman Drake* (Dordrecht: Kluwer Academic Publishers, 1990), 165–95.

77. Carl Gottlieb Bretschneider, ed., *Philippi Melanthoniis Opera quae supersumt omina*, Corpus Reformatorum, vol. I (Halle: C.A. Schwetschke, 1834), 42–46. A translation of Melanchthon's Preface can also be found in Robert Hand's edition of *On The Judgements of the Nativities* (Reston, VA: Arthat Press, 2001).

78. Robert S. Westman, "The Melanchthon Circle, Rheticus, and the Wittenberg Interpretation of the Copernican Theory" (Washington, DC: Smithsonian Institution, 1975).

79. Ibid, 179.

80. William Hammer, "Melanchthon, Inspirer of the Study of Astronomy; with a Translation of his Oration in Praise of Astronomy," *Popular Astronomy* 59 (1951): 308–19.

81. Melanchthon's paraphrase of Genesis 1:14.

82. For more on Pico and the practice of astrology in the late fifteenth and early sixteenth centuries see the seminal

study of Steven Vanden Broecke, *The Limits of Influence: Pico, Louvain and the Crisis of Renaissance Astrology* (Leiden: E. J. Brill, 2003).

83. Fantoni's manuscript, Conventii Soppressi B.7.749 is at the Biblioteca Centrale in Florence and contains 118 lectures on the *Tetrabiblos*. For more on this, see Rutkin, "The Use and Abuse of Ptolemy's Tetrabiblos," 142.

84. N. W. Swerdlow, "Johannes Petreius' Letter to Rheticus," *Isis* 82 (1992): 270–74. The book mentioned here is not Schöner's book on the Nativities but rather a manuscript by the fourteenth-century physician Antonius de Montulmo, entitled *De iudiciis nativitatum*. The earlier 'Arabic' book mentioned in the letter is the *Liber genethliacus, sive De nativitatibus* by Albubatris (Abu Bakr ibn al-Khasil al-Karachi, 10th century).

85. Claudia Kren, "Planetary Latitudes, the Theorica Gerardi and Regiomontanus," *Isis* 68 (1977): 194–205.

86. *Tetrabiblos*, IV.10.

87. All quotations from *De revolutionibus* come from Nicolaus Copernicus, *On the Revolutions*, translated by Edward Rosen (Baltimore: Johns Hopkins University Press, 1992).

88. See Noel M. Swerdlow, "The Derivation and First Draft of Copernicus' Planetary Theory: A Translation of the Commentariolus with Commentary," *Proceedings of the American Philosophical Society* 117 (1973): 423–512.

89. Ibid, 426.

90. Quoted from Edward Rosen, *Nicolas Copernicus: Minor Works* (Baltimore: Johns Hopkins University Press, 1985), 75.

91. Burmeister, *Georg Joachim Rhetikus*, Vol III, p. 50.

92. Edward Rosen, *Three Copernican Treatises* (New York: Columbia University Press, 1939), 109.

93. Burmeister, *Georg Joachim Rhetikus*, Vol III, p. 50 and Westman, "The Melanchthon Circle," 183.

94. Rosen, *Three Copernican Treatises*, 5.

95. Ibid., 111.

96. Nicolaus Copernicus, *On the Revolutions*, trans. Rosen, 285.

97. Rosen, *Three Copernican Treatises*, 109.

98. Zinner compared the paleography found on the Tusi couple with ÖNB, Codex 5002.

99. Owen Gingerich, "A Tusi Couple from Schöner's De Revolutionibus?," *Journal of the History of Astronomy* 15 (1984): 128–133.

100. Bas C. van Fraassen, *The Scientific Image* (Oxford: Oxford University Press, 1980).

101. Pierre Duhem, *To Save the Phenomena: An Essay on the Idea of Physical Theory from Plato to Galileo* [1908], trans. Edmund Doland and Chaninah Maschler (Chicago: University of Chicago Press, 1969), 56–57.

INDEX

Page numbers in **bold** indicate illustrations